KIRSTENBOSCH
the most beautiful garden in Africa

*Dedicated to the visionary founders of this national
treasure, to all who have worked in it, and to all
who love and protect the flora of South Africa*

Yellow cultivar of Aloe arborescens *'Philip le Roux'*

Published by Struik Nature
(an imprint of Random House Struik (Pty) Ltd)
Reg. No. 1966/003153/07
Wembley Square No. 2, First Floor, Solan Road,
Gardens, Cape Town, 8001
PO Box 1144, Cape Town, 8000 South Africa

Visit **www.randomstruik.co.za** and join the Struik Nature Club for updates, news, events and special offers.

First published in 2012
3 5 7 9 10 8 6 4 2

Copyright © in text, 2012: Brian J. Huntley
Copyright © in photographs, 2012: as indicated alongside images; images not attributed are drawn from SANBI archives or the public domain
Copyright © in maps, 2012: SANBI and Random House Struik (Pty) Ltd
Copyright © in published edition, 2012: Random House Struik (Pty) Ltd

Publisher: Pippa Parker
Editor: Helen de Villiers
Designer: Janice Evans
Cartographer: James Berrangé
Proofreader: Lesley Hay-Whitton
Indexer: Cora Ovens

Reproduction by Hirt & Carter Cape (Pty) Ltd
Printed and bound by
Tien Wah Press (Pte) Ltd, Singapore

All rights reserved. No part of this publication may be reproduced, stored in a retrieval system, or transmitted, in any form or by any means, electronic, mechanical, photocopying, recording or otherwise, without the prior written permission of the copyright owner(s).

ISBN 978 1 43170 117 9
Boxed edition 978 1 77584 003 9
Leatherbound edition 978 1 77584 004 6

Published with the support of the
South African National Biodiversity Institute
and the Botanical Society of South Africa

Contents

Foreword 6

Preface 7

Acknowledgements 8

Introduction – The Kirstenbosch story 10

CHAPTER ONE 14
'This is the place' – Harold Pearson
and the founding of Kirstenbosch

CHAPTER TWO 34
Developing the Garden –
progress from concept to reality

CHAPTER THREE 56
The Cape Floral Kingdom – The world's
hottest 'hot spot' of botanical diversity

CHAPTER FOUR 74
The plant hunters – Discovering and
documenting the diversity of life

CHAPTER FIVE 96
New directions in a changing South Africa –
The National Botanical Institute years

CHAPTER SIX 108
A Garden for all seasons –
gardens within the Garden

CHAPTER SEVEN 134
Conservation through cultivation –
An integrated approach

CHAPTER EIGHT 150
Come smell the flowers –
Inspiration and education

CHAPTER NINE 164
Conservation science –
Understanding the workings of nature

CHAPTER TEN 184
From poverty to prosperity –
Achieving financial sustainability

CHAPTER ELEVEN 206
A network of National Botanical Gardens –
Rolling out the dream

Bibliography 236

Index 237

Foreword

For a century, Kirstenbosch National Botanical Garden has served as the principal lens through which the world has viewed – and marvelled at – the incredible botanical riches of southern Africa.

Commencing some 16 million years ago, the initiation of glaciation in Antarctica and its subsequent split from South America led to the formation of the cold Benguela Current, flowing up Africa's west coast. This current cooled the prevailing westerlies as they passed onto warm land, limiting the amount of rain falling in the summer months. The resulting climate provided a setting for the evolution of the most bountiful radiation of botanical glory found anywhere on this planet. Four centuries ago these plants had already enriched the gardens of Europe, exciting the public imagination about the seemingly limitless array of novel floral beauties that they represented.

When in 1486 Portuguese sailors rounded the Cape and were enabled to pass eastward in the pursuit of their relentless search for the riches of India and tropical Asia, they named their inflection point 'The Cape of Good Hope'. For them it was hope of gaining untold wealth to the east, but it presents a different kind of symbolism for the modern world. Having lived under the increasingly oppressive system of apartheid for nearly half a century, contemporary South Africa has become a test case for racial integration, building the kind of diverse society on which the future of the world ultimately rests. This elegantly written and sumptuously illustrated book reflects the positive and innovative spirit that prevails in this extraordinarily interesting country.

Kirstenbosch has long presented the stunning natural beauty of the nation in a magnificent setting. As the crown jewel of the Nation's impressive system of National Botanical Gardens, Kirstenbosch is now serving all of the nation's people well and at the same time providing a wonderful showcase for visitors. We all take pride in its century-long record, joining in the celebration of what these gardens have meant in the past and, more importantly, what they are so well poised to contribute in the future.

PETER H. RAVEN
President Emeritus
Missouri Botanical Garden,
St Louis, United States of America

Haemanthus coccineus, *introduced to Europe in 1605.*
Watercolour painting by Vicki Thomas

Preface

Kirstenbosch is South Africa's flagship botanical garden. Despite its international fame and local popularity, there has not been a comprehensive account of its history and progress, ever since Compton's 1965 *Kirstenbosch, Garden for a Nation* went out of print many decades ago. Subsequent publications, such as Rycroft's 1980 *Kirstenbosch* and McCracken & McCracken's 1988 *The Way to Kirstenbosch* have been valuable additions to the literature, but lack coverage of the important developments of the past two decades. The approach of the Garden's centenary in 2013 signalled the need for a fresh account of the Kirstenbosch story.

This book draws heavily on information included in the annual reports published regularly since 1914 by the National Botanic Gardens of South Africa, by its successor, the National Botanical Institute, from 1990, and from 2004 by the South African National Biodiversity Institute. Much has been written on Kirstenbosch in the *Journal of the South African Botanical Society* and in the popular publication of the Botanical Society – *Veld & Flora*. Other important sources include Lighton's 1973 *Cape Floral Kingdom*, Gunn & Codd's 1981 *Botanical Exploration of Southern Africa*, and Karsten's 1951 *The Old Company's Garden at the Cape and its Superintendents*. Mike and Liz Frazer's 2011 beautifully illustrated *The Smallest Kingdom* is a wonderful addition to the literature on the topic. These and other key sources are listed in the bibliography. Because this book is a celebration of Kirstenbosch rather than a scientific text, detailed footnotes and citations of specific literature references are not included.

A clarification of the use of certain terms is appropriate. During the Victorian era, the many gardens established throughout the British Empire were named 'botanic' gardens, such as Royal Botanic Gardens, Kew, Durban Botanic Garden, New York Botanic Garden and many others. The original form of their names has been retained, as have references in the text to the literature of that time. However, the modern English usage of 'botanical' garden has been adopted by most gardens established during the 20th century.

The history of Kirstenbosch, of botany and of botanical gardens in South Africa covers a period of several centuries. Many remarkable individuals have contributed to this history, and, where helpful, their dates of birth and death are indicated to provide points of reference within the extended period covered by this narrative.

Welwitschia mirabilis. Watercolour painting by Barbara Pike, Kirstenbosch Biennale 2004

Acknowledgements

The history of Kirstenbosch is characterised by the passionate support and generosity it has received from the many people who love the Garden. During my 19 years of living in and working for the Garden, I enjoyed continuous inspiration and encouragement from colleagues, friends and family, and from many key role-players in the corporate, political and academic sectors.

Members of the Board of Trustees of the then National Botanical Institute, especially M.C. Botha, Kay Bergh, Saliem Fakir, Colin Johnson, Keith Kirsten, Gerhard Krone, Owen Lewis, Abri Meiring, Stuart Saunders, Jan Steyn, Ton Vosloo and Werner Zybrands, were generous in their support and guidance, and consistent in championing new ideas and innovations and in sharing risks.

In implementing the Kirstenbosch Development Campaign, Daan Botha, then Director for Gardens, with the support of our architects – Julian Elliot, David van den Heever and David Lewis – ensured that the Garden was provided with a lasting legacy of fine infrastructure. Kirstenbosch Curators John Winter and Philip le Roux and their excellent teams of specialist horticulturists, groundsmen and support staff have demonstrated their collective skills and dedication to the Garden in the quality and diversity of floral display, estate management and visitor amenities.

Christopher Willis, as the driving force behind the recent infrastructural development of the country's unique network of National Botanical Gardens, has shared much of the pain and pleasure of raising funds, overseeing capital developments and mentoring the new generation of curators who are responsible for the network's success.

Kirstenbosch has comprehensive modern facilities due very largely to the great generosity of many private individuals who have made donations towards development projects. Leading the Kirstenbosch Development Campaign through its most active years was Kay Bergh, supported by the Botanical Society, the foundational partner to which the Garden owes so much. Many Botanical Society members have made generous financial contributions to the Garden, a tradition that started at the time of its establishment. Some of these persons are mentioned in the text, but the list is too long to give all donors adequate acknowledgement

here. The major contributions of Mary Mullins, Kay Bergh, Ian Reddihough, Allan Bird and Leslie Hill must go on record. Similarly, the many corporations and foundations that have invested in Kirstenbosch deserve our thanks. These include the Anglo American and De Beers Chairman's Fund, the Rufford Maurice Laing Foundation, WWF South Africa, the Table Mountain Fund, the Rowland and Leta Hill Trust, the Molteno Brothers' Trust, Old Mutual, Sanlam, Appletiser and Pam Golding Properties. The Department of Environmental Affairs and Tourism also provided support, often generously.

Advice and guidance during the preparation of this book came from many colleagues, both within SANBI and beyond. Drafts of the text were rigorously reviewed by Jane Carruthers, John Donaldson, Graham Duncan, Roleen Ellman, Anthony Hitchcock, Philip le Roux, Dot Malan, John Manning, Eugene Moll, Ted Oliver, Tony Rebelo, Michael Rutherford, Ernst van Jaarsveld and Werner Voigt. The meticulous editing of advanced drafts by Otto Leistner and Christopher Willis is especially appreciated. Any errors or lapses that remain are mine.

The photographic material that enriches this book has been contributed by many SANBI staff, and friends of the Garden. My thanks are especially due to Adam Harrower, Specialist Horticulturist at Kirstenbosch, who dedicated many hours to photographing the plants, scenes and moods that portray the essence of the Garden, and contributed the majority of the images. Other members of the SANBI team, in Kirstenbosch and the other eight NBGs, generously made their photographic collections available for use in the book – Alice Notten, Anthony Hitchcock, Graham Duncan, John Manning, Ernst van Jaarsveld, Christopher Willis and Werner Voigt. Andrew Jacobs and Alice Notten at Kirstenbosch assisted in sourcing early photographs, and Anne-Lise Fourie at the Mary Gunn Library kindly helped track down some early publications. The tireless and ever patient support of Di Stafford and Gigi Laidler, my personal assistants during many years at Kirstenbosch, deserves warm acknowledgement. David Davidson is thanked for the beautifully delicate endpaper design.

The production team at Struik Nature, Pippa Parker, Helen de Villiers and Janice Evans, have been a pleasure to work with – demanding, charming and highly professional.

Finally, I must thank my wife, Merle, for sharing discussions with me at every step of the planning and development of projects at Kirstenbosch during our 19 years in the Garden. As an omniscient observer of all the complex and at times stressful processes of initiating and implementing the Kirstenbosch Development Campaign, and as hostess to the countless fund-raising functions that contributed to the success of the programme, Merle deserves a special mention of thanks.

BRIAN J. HUNTLEY

Otter Close, Betty's Bay,
Kogelberg Biosphere Reserve

Introduction

The Kirstenbosch story

The Kirstenbosch story has been frequently related. It is a tale of conflicts between Dutch settlers and Khoikhoi pastoralists in the 17th century, of failed attempts to protect the forests of the mountain ravines during the 18th century and of the struggles of the Cape's leading botanists to establish a botanical garden on the site of the old Company Garden during the mid-19th century. But the story of Kirstenbosch really begins with the dynamic leadership of Harold Pearson, the person who so comprehensively articulated a vision for a modern, uniquely South African institution during the early days of the 20th century.

A rage of colour in spring, the mesembryanthemum banks blaze with red and orange forms of Drosanthemum speciosum, *yellow* Drosanthemum bicolor *and, in the distance, purple* Lampranthus amoenus. *Castle Rock provides an iconic backdrop to the Garden.*

Henry Harold Welch Pearson's vision, spelt out in his address to the South African Association for the Advancement of Science in 1910, provides a lasting framework for what developed into an institution of global significance. His thinking, and that of the many great personalities that guided the development of the botanical sciences and, ultimately, the biodiversity conservation agenda of the country, underpins much of the Kirstenbosch story as presented here.

This narrative begins with an account of the fortuitous selection of the site of the new garden – 'this is the place'. The decision, once made, was never regretted. The excellent choice of site was followed by superb choices in the leadership of the project – Pearson, Mathews, Compton: all professionals of the highest quality, supported by the most eminent figures in politics, business and society. The early years of the Garden, described in chapter 2, reflect the sociopolitical construct of the time, building on a colonial, Victorian legacy, but also moving forward to the ultimate incorporation of the developmental agenda of post-1994 South Africa.

The distinctive setting of Kirstenbosch, at the epicentre of the Cape Floral Kingdom, and in one of the world's 'hottest hot spots' of biodiversity, provided the Garden's champions with a unique opportunity to present to the public the spectacular beauty and diversity of our flora. This richness of species, its evolutionary origins, and the challenges of threats to its survival in a rapidly changing landscape, are described in chapter 3.

The Cape of Good Hope emerged during the late 18th century as the destination of choice among botanical explorers of the age. The long tradition of botanical exploration by these intrepid men of the Enlightenment – 'searchers after truth' – has been fundamental to the strength of South Africa's botanical sciences. Chapter 4 provides a brief account of the heroes of South African botany, from the 17th century to the present day.

The dramatic sociopolitical changes resulting from the advent of democracy in South Africa in 1994 introduced new opportunities for change in Kirstenbosch. Chapter 5 describes the impetus given to the Garden's development by a new 'business model', a new political landscape, and the role of partnerships in securing the funding needed for major capital developments.

In chapters 6 and 7, the core business of Kirstenbosch is presented – plant display, horticulture, conservation, and landscape design. The special environmental conditions of the Garden's position on the east-facing slopes of Table Mountain, its soils, microclimate and topography, all offered the opportunity for a Garden of stunning visual impact. The richness and vulnerability of the Cape Floral Kingdom give the Garden's skilled horticulturists and botanists unusual technical and intellectual challenges, met with great success – and occasional failure.

From its earliest days, Kirstenbosch has played a central and leading role in the development of environmental education through the use of the Garden as an outdoor classroom. This has expanded into a broad-based 'outreach' programme, taking the Garden to the community, most especially to those less privileged than the traditional Kirstenbosch visitors. Chapter 8 introduces the role of the Garden in inspiring and exciting a new generation of environmentally responsible citizens.

Botanical gardens are, by definition, centres of research on aspects of the flora, vegetation, horticulture and ecology of plants. Until the 1980s, the research programme at Kirstenbosch focused almost entirely on taxonomic studies – collecting, identifying, classifying and archiving plants of the Cape Floral Kingdom. Chapter 9 describes how the science programme at Kirstenbosch has rapidly expanded, in response to global trends and society's needs, to include studies on climate change, threatened species, land-use policy and, since the establishment of SANBI, the faunal component of South African ecosystems – both terrestrial and marine.

Much of the literature on botanical gardens relates to the financial trials and tribulations of garden directors. Perceived poverty is a subtext of much of the history of South Africa's gardens. Chapter 10 demonstrates a different story – the transition of Kirstenbosch's fortunes from poverty to prosperity. Using the multiple opportunities offered by the Garden's great intrinsic assets – location, grandeur, popularity, accessibility – and the strategic investments in visitor, education, research and horticultural facilities, the success of the Kirstenbosch 'business model' is presented.

The final chapter of the Kirstenbosch story is that of the concept and development of a network of National Botanical Gardens (NBGs) of South Africa. From humble beginnings at Matjiesfontein to the present network of nine NBGs, Pearson's vision of a national institution embracing the full diversity of the country's unparalleled floral richness has been achieved.

The development of Kirstenbosch started in 1913 with the cycad amphitheatre, in the heart of the Garden. Fernwood Buttress rises to over 1 000 metres above the Garden, behind which the highest point on Table Mountain, Maclear's Beacon, at 1 085 metres, marks the far northeastern corner of the Upper Kirstenbosch Nature Reserve, now part of Table Mountain National Park.

Introduction

This is the vista of Kirstenbosch, viewed from the upper slopes of the Garden, that so inspired William Burchell on his visit in March 1811.

CHAPTER ONE

'This is the place'

Harold Pearson and the founding of Kirstenbosch

Our walk conducted us to a high point of the hill that overlooks Kirstenbosch, a beautiful estate belonging to the Government ... The view from this spot and indeed all the scenery around, is the most picturesque of any I had seen in the vicinity of Cape Town. The beauty here displayed to the eye could scarcely be represented by the most skilful pencil.

WILLIAM J. BURCHELL
Travels in the Interior of South Africa, *1822*

'This is the place'

The remarkable William Burchell, who visited the Cape from November 1810 to August 1815, was arguably the most observant, productive and articulate traveller to visit South Africa in the early 19th century. His botanical, zoological and geological collections numbered over 63 000 items, many new to science. His notes on the people he met during his lonely wanderings over four years and 7 000 kilometres demonstrate great sensitivity and compassion. His ecological understanding of the workings of the South African veld was a century ahead of his time. His description of Kirstenbosch, made after his visit on 14 March 1811, is as true today as it was 200 years ago.

In their search on 10 February 1911 for a site for a new National Botanic Garden, Pearson, Pillans and Ridley would have encountered the somewhat abandoned Kirstenbosch estate recorded in this photo taken in 1905. The cottage was to become home to J.W. Mathews and successive curators until 1947.

ABOVE *Silver Trees and the view towards Devil's Peak, from the site of Burchell's view, painted by Gwelo Goodman in 1932 (Courtesy of Camilla Kuiper)*

LEFT *William Burchell (1781–1863) was unquestionably the most prolific collector of botanical, zoological and geological material in the history of South Africa.*

The story of Kirstenbosch commences a century after Burchell's visit. On the sunny afternoon of 10 February 1911, Henry Harold Welch Pearson, Professor of Botany at the South African College, together with a young botanist, Neville Pillans, and George Herbert Ridley, Curator of the Cape Town Municipal Gardens, set out on a horse-drawn Cape cart to look for a site for a new botanical garden. Pearson had his mind set on some land on the wooded slopes south of the Groote Schuur zoo. Young Pillans had clearly explored more widely, or perhaps remembered reading Burchell's comments. Whatever the case, the trio continued onwards and past the Bishopscourt homestead, and up the lane that ended at the avenue of Moreton Bay Figs and Camphor Trees, planted by Cecil John Rhodes in 1898.

Emerging from the grove of young trees, they reached the site that became, and remains, the iconic point of entry into Kirstenbosch – the verdant sweep rising to the grandeur of Castle Rock, flanked by the rugged, forested eastern face of the Table Mountain massif. Without hesitation, Pearson, overwhelmed by the majesty of the prospect, exclaimed 'This is the place'. The rest is history.

We will return to Pearson, his vision and his impact (see page 27). But first we need to look back a few centuries – to the history of Kirstenbosch.

Early days – from woodsman to empire builder

Robert Compton, the longest-serving Director of Kirstenbosch (1919–1953), in his charming and detailed book *Kirstenbosch, Garden for a Nation* observes that the story of Kirstenbosch is 'compounded of romance and hard facts'. To this one might hasten to add 'passion and hard work'. But there is much that is romantic about the place, and no shortage of urban legend, even myth. To appreciate the story, we must look back to its early beginnings.

The written history of Kirstenbosch starts in 1657, and has been carefully researched by Mary Alexander Cook, who collaborated with Compton on his book on the Garden. She found that the first record of the land that was to become Kirstenbosch was in the granting of woodcutting rights to the forests on the east-facing slopes of Table Mountain. On 27 October 1657, Leendert Cornelissen of Zevenhuysen, a free carpenter and sawyer, previously in the employ of the Dutch East India Company, obtained the right to utilise the forests. In the company of Commander Jan van Riebeeck, Cornelissen had visited the forests the previous day – a good three hours travelling from Cape Town. Cornelissen's responsibility, besides cutting and sawing timber for sale in the rapidly growing town, was to prevent indiscriminate depredations on the forests. During the 17th and 18th centuries, the forests became known as 'Leendertsbos'.

Isolated outliers of the extensive Afrotemperate Forests of Africa approach their southern limit on the slopes of Table Mountain. This remnant of the once-dense forests of 'Leendertsbos' shades the Silver Tree Trail along the Lower Contour Path.

The forest protection responsibilities included in Cornelissen's grant established the first formal conservation measure taken in the history of South Africa – an auspicious start for what was destined to become the country's flagship of biodiversity conservation. Van Riebeeck went further by formalising protection for forests, and especially the Real Yellowwood *Podocarpus latifolius*, through the proclamation of a *placaat* on 12 October 1658, prohibiting the cutting of Yellowwood trees 'no matter under what pretext'.

A sign of the times was that, as additional conditions for receiving the grant, Cornelissen had to remain a freeman in the Cape for 15 years, and should 'order his wife from the Fatherland'. Kirstenbosch was remote from the settlement in Cape Town, lonely and dangerous, and it required tough, indeed rough, personalities. Cornelissen started well: he treated his slaves 'the best of all the freemen', and on 1 May 1660 he was nominated for the office of 'burger councillor'. But his status was short-lived; on 25 October 1661 it was noted that:

> *'The free sawyer Leendert Cornelissen of Zevenhuysen, a burger councillor elected last year, instead of setting the freemen an honourable and dignified example, has daily been behaving in a more and more debauched manner, by drinking, celebrating, fighting, brawling, swearing, etc.... it was resolved to discharge him from his office as burgher councillor'.*

So ended the career of Kirstenbosch's first 'curator'.

Van Riebeeck's hedge

Another first for Kirstenbosch, embedded in the sad history of the country, was the planting, in 1660, of a Wild Almond *Brabejum stellatifolium* hedge from the slopes of Table Mountain, along Wynberg Hill, to the 'twisted tree' near Groote Schuur. The stated purpose of the hedge was to curtail the activities of cattle-raiding Khoikhoi by the establishment of a boundary fence and live hedges to demarcate the settlement's thirteen-and-a-half-kilometre periphery. Later historians would express a different view – the Khoikhoi were merely claiming grazing and watering rights denied them by the colonists. To some, the dense, impenetrable hedge was no less than the first manifestation of what became the apartheid policy of the 20th century.

It has been argued that the Wild Almond hedge was a failure, and that the thickets of this species today recognised as 'Van Riebeeck's Hedge' and proclaimed a National Monument in 1936 are simply remnants of natural populations of Wild Almond growing spontaneously along the ridge. This is improbable, as Wild Almond trees invariably grow along streams and rivers – not on dry ridges. To deny the original stated purpose would remove both historical fact and an element of romance from the story.

'Van Riebeeck's Hedge' was proclaimed a National Monument in 1936.

Many years passed before the forests of Leendertsbos were again mentioned in the annals of the Cape. This was not until 1795, when the British occupied the Cape for the first time. An inventory of the Dutch East India Company's property was drawn up. What had been known as Leendertsbos was now referred to as Kirstenbosch. Where did this new name come from?

Much has been written on the topic. The most probable answer is that it was named after a member of the Kirsten family, who owned land in several areas nearby. Willem Hendrik Kirsten might have been post-holder of the property at some time in the late 1780s. Other explanations suggest that a Cherry orchard (*Kersebos*), planted behind the old Company house built on the property in the 18th century, or the abundance of Kersbos *Pterocelastrus tricuspidatus* in the area, might have given rise to the name. Whatever the case, the origin of the name is lost in the history of the Garden. May it so remain!

Following the turbulent events in Europe at the turn of the 18th century, the Cape was governed successively by the Netherlands, Britain, the Batavian Republic and, from 1806 again, Britain. The cost of administering the colony was significant and, to raise funds, Britain started selling off lands previously owned by the Dutch East India Company. Among the first of the properties to be granted freehold was Kirstenbosch – two units of which, each of about 92 hectares, were sold to the Deputy Colonial Secretary Colonel Christopher Bird, and to the Colonial Secretary Henry Alexander.

Wild Almond Brabejum stellatifolium *provides a robust, contorted and impenetrable barrier to the passage of man or beast – making it ideally suited to the purpose for which it was planted in 1660.*

This is the place 19

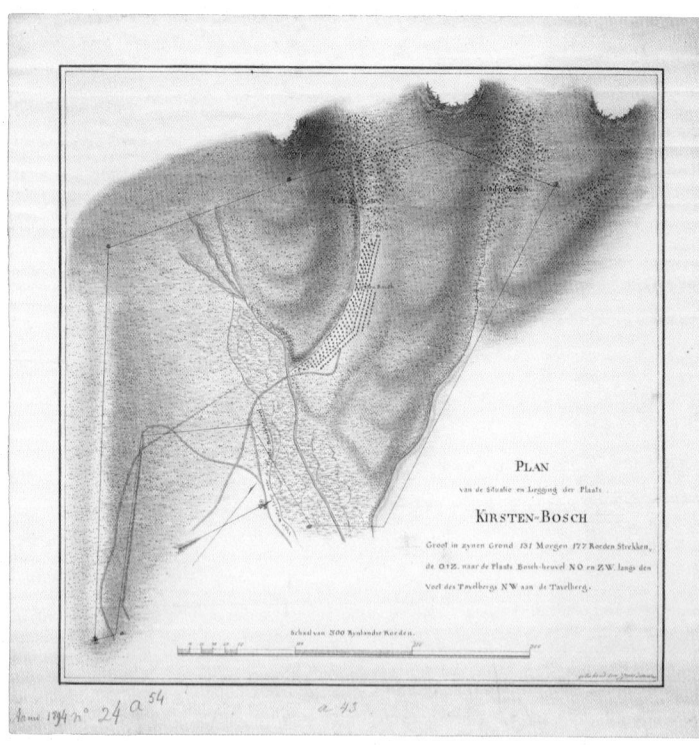

ABOVE *A plan of Kirstenbosch drawn up between 1802 and 1806 reveals rows of orchard plantings, just discernible in the heart of the Garden. (J.W. Janssens 1748–1825, courtesy of Nationaal Archief, Den Haag)*

Colonel Bird soon sold his land to Alexander, but not before building a small pool in the shape of a bird at the natural spring in the heart of the property – the delightful 'Bird's Bath' that has remained, two centuries later, a focal point in the Garden. Romantics have called the pool 'Lady Anne Barnard's Bath' – conjuring visions of the early 19th century beauty taking her toilet in the Arcadian dell – but she had already left the Cape before Bird took ownership of the land, so the story remains but a charming myth.

Alexander built a house directly below Castle Rock, today occupied by the Marquee Lawn, and some few hundred metres below the old Dutch East India Company house, already derelict at the time. Alexander died in 1818 and his land reverted to the Colonial government. Kirstenbosch changed hands several times through the 19th century: in 1823 from the Company to Dirk Gysbert Eksteen; in 1853 through marriage to Charles Duffy Henry Cloete; and in 1895 to Cecil John Rhodes. Rhodes had the good sense to use his considerable wealth to purchase as much land along the eastern slopes of Table Mountain as he could, to protect the land from urban development. He paid £9 000 for the derelict 130-hectare farm. Along with much other land, Kirstenbosch was left by Rhodes to 'the united peoples of

RIGHT *The homestead built by Eksteen in the 1820s on the site of Henry Alexander's house was extended by Cloete in the 1850s. Surviving until the first years of the Garden's development, it was demolished in the 1920s and the site was subsequently occupied by the first Tea House, and most recently developed as a marquee lawn.*

South Africa' when he died in 1902. His vision of a united country became a reality with the establishment of the Union of South Africa on 31 May 1910. The property was transferred to the Union government on 11 February 1911.

Kirstenbosch's good fortune was that soon after Rhodes' death in 1902, a similarly ambitious man of vision arrived in Cape Town – Henry Harold Welch Pearson (1870–1916). While Rhodes' energy was driven by mineral wealth and empire building, Pearson was inspired by the country's living assets – its abundant and beautiful flora. Pearson had no sooner settled into life in Cape Town than he began promoting the idea of a National Botanic Garden for the future Union of South Africa. But before we describe Pearson's campaign, we need to consider earlier initiatives.

In 1895 Cecil John Rhodes purchased Kirstenbosch and other properties along the eastern slopes of Table Mountain. Bequeathed to 'the united peoples of South Africa' in his will, the land was used to develop Kirstenbosch, the University of Cape Town and Groote Schuur Hospital.

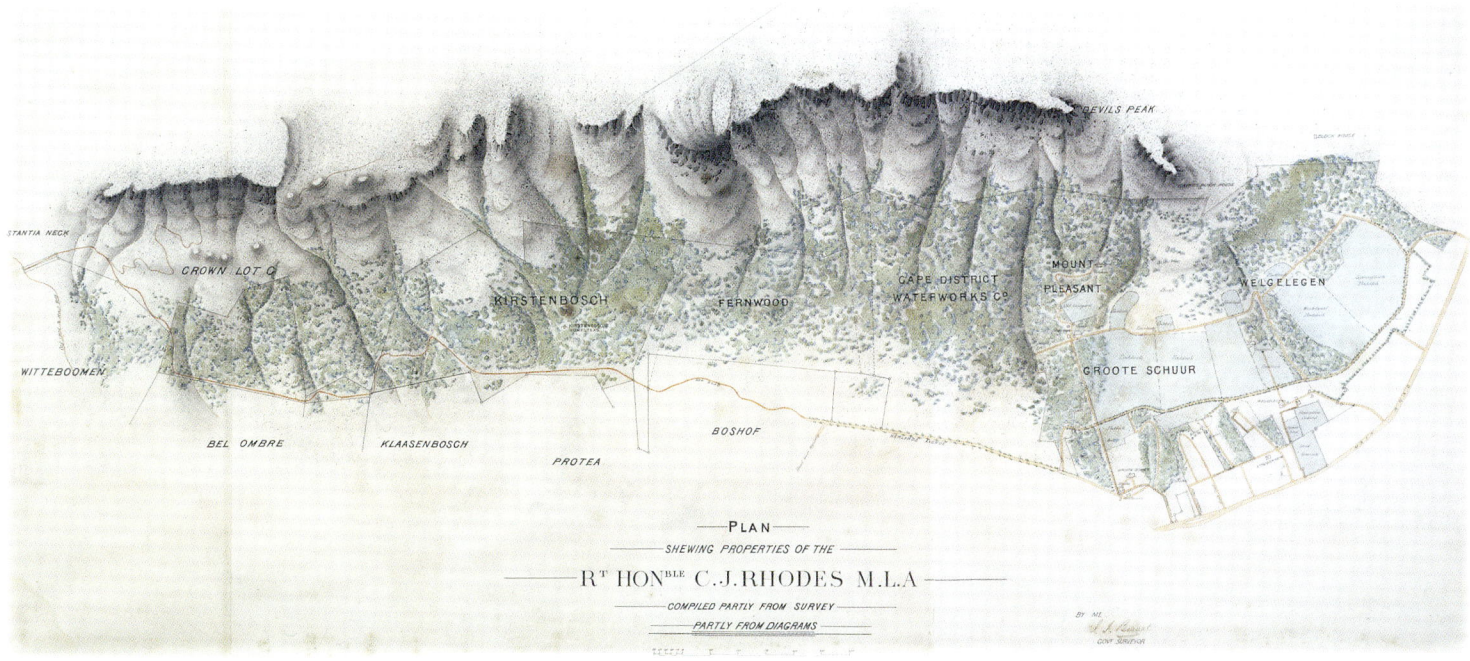

This map outlines Rhodes' extensive properties along the eastern slopes of Table Mountain.

A recent photo shows the eastern slopes of Table Mountain – the backdrop to what was Rhodes' estate, now the site of Kirstenbosch.

This is the place

The east-facing slopes rising above the Garden are clothed in fynbos. Old-growth fynbos, not burnt in 40 years, covers the slopes below Castle Rock, while that to the left was burnt in 2002. Afrotemperate Forest has regenerated in the ravines, where the forests were first exploited by Leendert Cornelissen in the 1660s, infested by alien invasive trees and bramble during the 19th century, and cleared through the 20th century. Aloe arborescens and Euryops pectinatus occupy the foreground.

In 1898 Cecil Rhodes planted an avenue of Camphor Trees to provide shade during his rides from Groote Schuur to Constantia. These had grown into a fine avenue by 1913, although the road was still little more than a rough track. (Courtesy of the Elliot Collection, National Archives)

Today, the Camphor Avenue is no longer the main thoroughfare from the city to Constantia and Hout Bay, but a tranquil and lush path into the upper Garden.

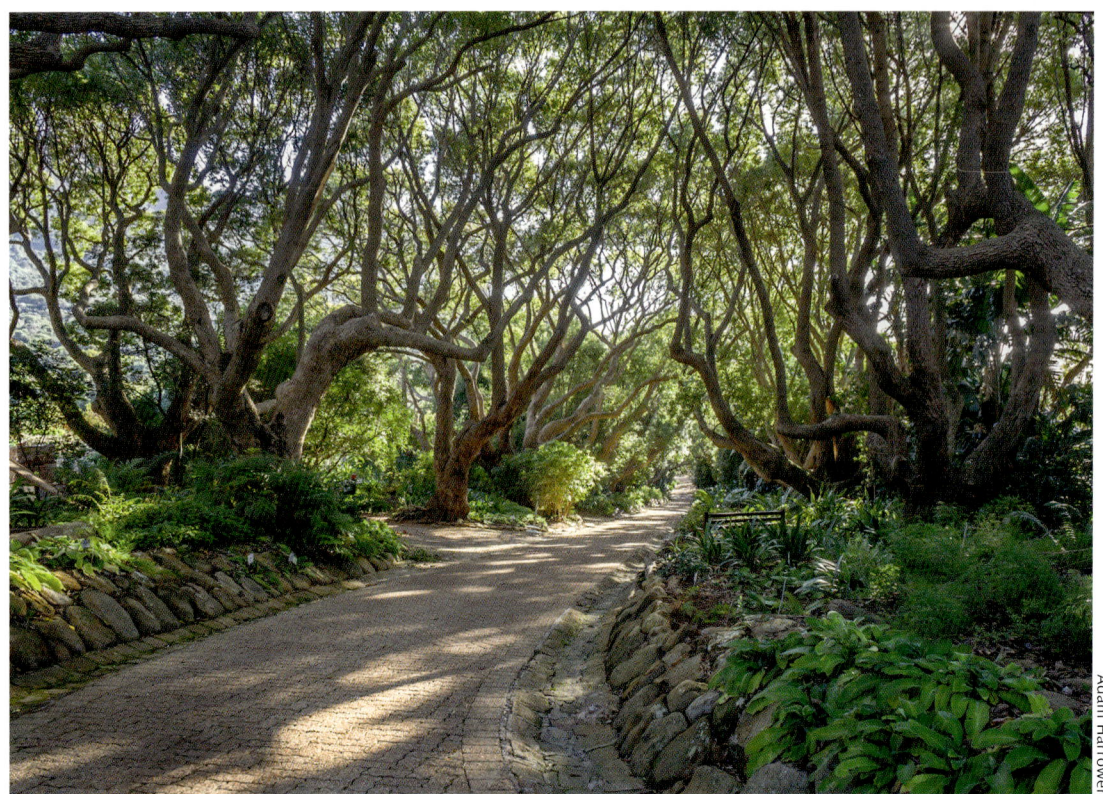

A botanic garden for Cape Town – the idea that failed

The idea of a well-equipped botanic garden in Cape Town had been mooted by many of Pearson's predecessors. First of these was the energetic William Burchell (1781–1863) who, after his extensive travels and highly productive collecting expeditions in South Africa from 1810 to 1815, continued his pioneering work in Brazil. While in Cape Town, he commented:

'If, in the vicinity of Cape Town, a well-ordered botanic garden, of sufficient extent, were established with the purpose of receiving plants which might casually, or even expressly, be collected in the more distant parts of the colony, the sum of money required for maintaining it would be but trifling in comparison with the advantages of which science, and the public botanic gardens of England, would derive from it.'

Cape Town already had the old Company's Garden, established by Jan van Riebeeck in 1652, but this was primarily devoted to growing vegetables and fruit both for the settlement and for provisioning the ships of the Dutch East India Company on their voyages to and from the East Indies. An early attempt to grow local, indigenous plants had been made by one of its superintendents, Johannes Andreas Auge, between 1747 and 1772, but there was no further systematic programme for growing Cape plants until Kirstenbosch was established in 1913.

By the 1830s the old, neglected Company's Garden was in 'a disgraceful state of abandonment'. Leading citizens petitioned for a new, real 'botanic garden', and in 1848 it was decided to establish such a garden in the city centre to replace the old Company's Garden. Leaders in the project included eminent botanists – among them Ludwig Pappe, the first Cape Colonial Botanist, appointed in 1858 and also holder of the first Chair of Botany at the South African College; Carl Zeyher, the great collector of southern African plants; and Peter MacOwan, one of the most articulate (and acerbic) botanists in the country, also Professor of Botany at the South African College. But despite the excellence of the key persons appointed to manage the Garden, shortage of funds and indifferent support from the Cape colonial government resulted in the project's failure. The site was small, lacking in water, and on poor soil. Petty squabbling between the Garden's commissioners, the government

Ludwig Pappe (1803–1862), first Colonial Botanist of the Cape

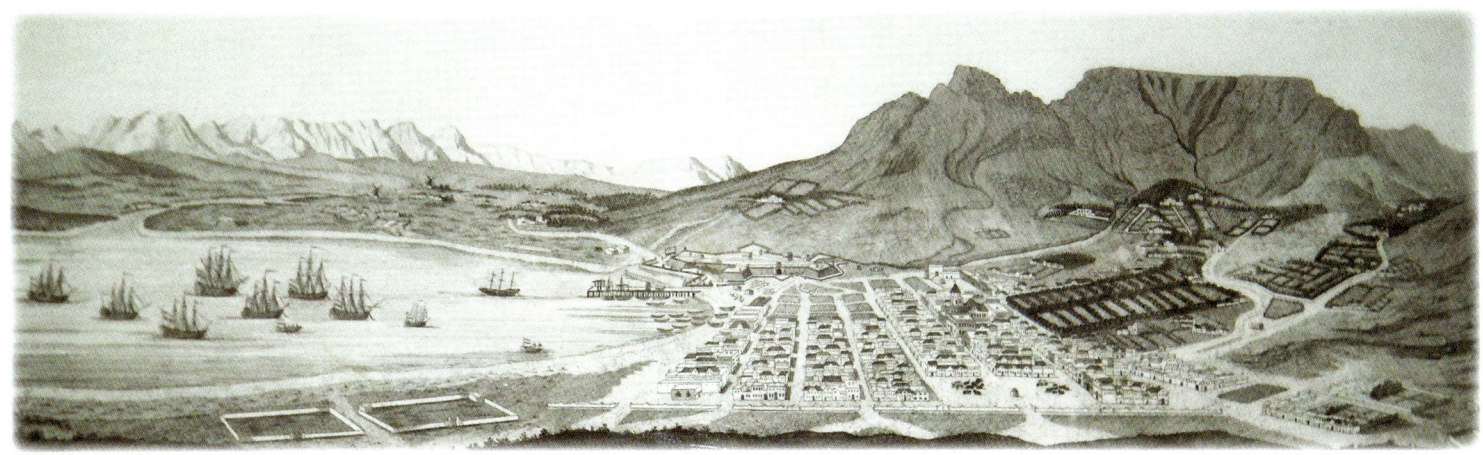

An early engraving (c. 1760s) of Cape Town shows the layout of the old Company's Garden built by Commander Jan van Riebeeck in 1652. (Courtesy of the Campbell Collections, University of KwaZulu-Natal)

This is the place

and the public characterised the Garden's half-century life. Negative commentary ranged from letters to the press, a parliamentary enquiry, and correspondence from the Director of Kew. In short, the Garden was a disaster. Harry Bolus, the doyen of botany in South Africa at the time, in giving evidence to a Parliamentary Select Committee in 1877, said 'I consider [the Garden] a great discredit not only to the town, but to the colony altogether'.

The Cape Town Botanic Garden fell short on all the criteria recognised as fundamental to any respected Botanic Garden in the 19th century: correctly labelled, living collections on public display, a well-organised herbarium, correspondence and exchange of material with other gardens, and strong financial support, all under the direction of a competent botanist.

The tombstone of Ludwig Pappe, the first Cape Colonial Botanist, is now set in the entrance wall of the Compton Herbarium.

Despite enormous difficulties, the Cape Town Botanic Garden's botanists did what they do best – collect plants. The collections of Pappe, Zeyher, MacOwan, and many others, were maintained in a poorly housed herbarium, but with the personal support of the then Governor, Sir Henry Barkly. The Cape Government Herbarium was eventually incorporated into the South African Museum Herbarium, which was started in 1828, and the collections were, in turn, transferred on loan in 1956 to Kirstenbosch and finally donated permanently to the Compton Herbarium, now home to the most important collection of type specimens in the country. Conversely, the Cape Town Botanic Garden was the point of entry to South Africa of many of the worst invasive alien species to reach our shores. The list of plants growing in the Garden in 1858 includes no fewer than 71 species of the most aggressive plant invaders to infest our Fynbos, Grassland and Forest biomes in years to come. The Garden also introduced the plant disease 'chrysanthemum rust' into the country.

In 1892 the Garden was transferred to the Cape Town Municipality. It was now known as the Cape Town Municipal Garden and, with increased funding and the abandonment of any pretensions of being a Botanic Garden, it served the people of Cape Town admirably as a municipal park. Well endowed with showy roses, dahlias, chrysanthemums and daffodils, it met its new and appropriate goal of providing 'the highest cultivation and presentation of plants and flowers with the greatest comfort and enjoyment of visitors'. This role, more than a century later, is sadly neglected.

Kirstenbosch – the vision that succeeded

Successive visitors to the Cape, local citizens and correspondents from abroad had, on repeated occasions, called for South Africa to establish a 'Kew in South Africa'. Throughout the history of Kirstenbosch, the influence of the Royal Botanic Gardens, Kew, has been profound, even if predictable, given Kew's pre-eminence in matters botanical. Thus it was that, when Harold Pearson left Kew to take up the Chair of Botany at the South African College in 1903, he would have had some well-founded opinions on the need for a Botanic Garden – as his former Director Thiselton-Dyer had expressed in 1895:

'At present the Cape Colony is the only important British Possession which does not possess a fully equipped Botanical Institution … A central establishment in the neighbourhood of Cape Town devoted to the scientific study and experimental cultivation of plants, fully equipped to discharge its duties as a national institution on the lines of Kew, would alone be worthy of the future of South Africa.'

Pearson's vision – a National Botanic Garden

Henry Harold Welch Pearson was born at Long Sutton, Lincolnshire, England on 28 January 1870. A brilliant student with a distinguished record at Cambridge, he won a travelling scholarship that took him to Ceylon (now Sri Lanka), to the famous Botanic Garden at Peradeniya, exposing him, at an early point in his career, to the challenges and possibilities of work in distant colonies. He returned to England for a spell as Assistant Curator of the herbarium at Cambridge University, and in 1899 he joined Kew as assistant to the Director, Sir William Thiselton-Dyer. Working with the dictatorial leader of this prestigious institution groomed him for his future role, not only as a scientist, but also as a strong-willed and determined negotiator among his academic and political peers.

In 1903 Pearson came to South Africa as the first Harry Bolus Professor of Botany at the South African College (from 1918, the University of Cape Town). His energy, charm and leadership resulted in rapid change. He built up the Botany Department from 'no students, no laboratories, no equipment' to a centre of excellence. His primary interest in botany focused on the gymnosperms – specifically the Gnetales – which include the enigmatic desert plant *Welwitschia mirabilis*. It is said that a principal reason for his accepting the post in Cape Town was his wish to study this plant *in situ*. His enthusiasm and energy took him far into the remote corners of southern Africa, including travels to the wilds of Angola. His research on the Gnetales earned him a Fellowship of the Royal Society in 1916, at the time – and still today – one of the highest accolades to which a scientist can aspire.

Without question, Pearson's signal contribution to South Africa was his clear understanding of what the establishment of a national botanical garden should entail, and how to achieve this. His vision of an institution that took advantage of the country's immense floristic wealth, and of the role of research, indigenous plant horticulture, economic botany and plant conservation, spelt out in his address to the South African Association for the Advancement of Science in 1910, provided a vision as relevant today as it was a century ago.

Henry Harold Welch Pearson (1870–1916) – founder of Kirstenbosch

Sir William Thiselton-Dyer, Director of the Royal Botanic Gardens, Kew, from 1885 to 1903, had a strong influence on Pearson, encouraging him to pursue a botanical career at the Cape. (Courtesy of Royal Botanic Gardens, Kew)

Pearson's primary botanical focus was on the gymnosperms, most particularly on the enigmatic desert plant Welwitschia mirabilis. *He trekked across southern Africa to see* Welwitschia *in southern Angola.*

This is the place

A garden focusing on indigenous plants would be an ideal living laboratory, especially if close to a university, allowing students to address the intriguing intellectual challenges presented by the extreme richness and high levels of endemism of the Cape flora. Pearson did not rush in to announce his ideas, however. He wisely spent his first several years in the Cape getting to know the country, its flora and, importantly, its politics. He built up a strong body of students, dramatically improved the teaching and laboratory facilities, and skilfully immersed himself and his wife in the influential circles of Cape Town society.

Pearson's scientific standing was soon recognised within the country. In 1910 he was elected President of the Life-sciences section of the South African Association for the Advancement of Science. On 10 November that year he delivered his presidential address on the topic 'A National Botanic Garden'. His paper was both welcome and provocative. It was welcome because he was stating what most of his peers, at least those in the Cape, had wanted to say, but lacked the scientific and international stature to argue convincingly. He called for a State-administered institution with a large garden, herbarium, museum and well-equipped laboratories, close to a university, and led by a competent botanist. He reinforced the opinion expressed many times previously that the garden should give special attention to the study and preservation of the indigenous flora of South Africa. He emphasised the need to study plants in their natural habitat, and to exploit their economic potential.

Pearson's campaign was provocative in that it exposed a certain naivety – perhaps deliberately? His emphasis on a *national* institution, based in Cape Town, from where all activities in a network of as many as 10 regional gardens would be co-ordinated, if not centralised, ignored the momentum that was building in the Transvaal for a national botanical survey and a national herbarium based in Pretoria. He was aware of the recent appointment of Dr Illtyd Buller Pole Evans (in 1905) to the Department of Agriculture in Pretoria, but perhaps did not anticipate the commonality of his own and Pole Evans' ambitions – something that was to cause his successor at Kirstenbosch, Compton, much distress. The criteria he proposed for site selection – which excluded a 'sub-tropical' site – seem to have been skilfully directed against the choice of Durban, which already had a fine Botanic Garden, established in 1851, and an excellent herbarium.

The difficult financial history of all South African colonial gardens no doubt influenced his emphasis on the relatively new subject of economic botany. 'The case for the establishment of a State Department of Botanical Research, with the Botanic garden as a fundamental part of its organisation, rests therefore upon an economic basis.' During his few years at Kirstenbosch he made valiant efforts to develop the economic potential of indigenous plants – a programme continued, with great frustration and ultimate failure, by his successor, Compton.

He recognised that the period immediately following the consolidation of the previously divided colonies and republics into the Union of South Africa was the political moment at which to present his vision. He was also aware of the confused state of botany in the country, with multiple and ever-changing, small, subcritical activities at local, regional and national levels. He played to the newly found nationalism of the Union of South Africa, identifying himself as a South African.

What made his vision of a new botanic garden so different from others around the globe was his emphasis on the study and preservation of the country's indigenous flora. Here we have what has become the unique selling point of Kirstenbosch (and that of other of South Africa's botanic gardens too). Whether this emphasis was his personal belief, or whether he was merely playing to his audience, we do not know. But what we do know is that he was an astute negotiator, and a flexible one at that. Having alarmed the then Prime Minister, General Louis Botha, with excessive cost estimates for the Garden (based on a simple extrapolation of costs of Peradeniya in Ceylon) of £32 000 per year, and of a possible duplication of duties with the existing programmes of the Department of Agriculture, he was quick to make compromise proposals. But on the location of the Garden in the Cape he did not budge.

Pearson had actively canvassed his vision for a new garden among the great and the good of Cape Town. Matters came to a head when a committee of distinguished citizens met on 8 March 1912 to carry forward the ideas expressed in Pearson's seminal papers of 1910 and 1911. The meeting was chaired by the eminent Lord Chief Justice, Baron De Villiers, and included the mining magnate and Member of the Legislative Assembly (MLA) Sir Lionel Phillips; the respected botanist Dr Rudolf Marloth; Senator W.P. Schreiner; John X. Merriman (MLA); the Mayor of Cape Town, Sir Frederick Smith and the Principal of the South African College, Dr J.C. Beattie, among other notables.

The meeting agreed that 'a committee be formed to consider the formation of a National Botanical Society to prepare details of a scheme for the establishment of a National Botanic Garden within the Peninsula, and to report to a subsequent meeting.'

With such influential support, Pearson was on the home run. He prepared material for a speech that Sir Lionel Phillips read in the House of Assembly on 6 May 1913. The hour-long address covered much ground – the suitability of Kirstenbosch as a site, its good soils, water supply, diverse vegetation and flora, and that it was already in government ownership. He moved '… that in the opinion of this House the Government should consider the advisability of setting aside a piece of ground at Kirstenbosch for the establishment of a National Botanic Garden.' His motion was supported with enthusiasm and carried unanimously.

Sir Lionel Phillips led a delegation that met with the Prime Minister on 27 May. General Botha gave them a sympathetic but cautious hearing: he agreed to setting aside Kirstenbosch for the Garden, but specified that the project should not duplicate the work of the Department of Agriculture. The government agreed to contribute £1 000 per annum for the running of the Garden, plus £2 500 to build a house for the director and for a small laboratory.

At a public meeting on 10 June, it was resolved to establish a National Botanical Society, with the specific objective of augmenting government grants towards the development of Kirstenbosch.

On 1 July 1913, Pearson's grand vision finally became a reality. Kirstenbosch was handed over to a statutory Board of Trustees under the chair of Lord De Villiers, with Sir David de Villiers Graaff and Sir Lionel Phillips as additional government representatives. The Mayor of Cape Town became an ex-officio member representing the city. After the formal establishment of the Botanical Society on 31 July, Mr W. Duncan Baxter was elected to represent it – a role that he played with great competence and devotion for nearly 45 years, critical years during which Kirstenbosch slowly came of age.

Predictably, the Board of Trustees appointed Harold Pearson as the first Honorary Director, for which he received no salary. He maintained his position as Harry Bolus Professor of Botany at the South African College. The link between the university (since 1918 the University of Cape Town) and Kirstenbosch continues to this day through the honorary Harold Pearson Chair of Botany created following Pearson's death.

Sir Lionel Phillips, Member of Parliament and patron of the arts, proposed the establishment of Kirstenbosch to the House of Assembly.

Louis Botha, as First Prime Minister of the Union of South Africa, approved the transfer of the farm Kirstenbosch to the Trustees of the National Botanic Gardens, conditional on the establishment of the Botanical Society of South Africa as a source of public support.

Baron De Villiers (1842–1914), Lord Chief Justice, was the first Chairman of the Board of Trustees of the National Botanic Gardens of South Africa.

This is the place

First steps – giving the Garden 'good bones'

Having won the battle to establish a new, truly South African National Botanic Garden, Pearson was now faced with the daunting task of building not only a new garden, but also a new institution. He had made great promises as to the nature of the new organisation – audacious proposals, given the financial, logistic and political dynamics of the time. But he clearly followed a philosophy of 'think big, start small' that has underpinned most of the evolving institution's successes.

Pearson's choice of right-hand man – Joseph 'Jimmy' William Mathews (1871–1949) – could not have been more fortunate. Mathews was a Kew-trained plantsman, who joined Kirstenbosch as Curator after several years in the Cape Town Municipal Gardens and as a commercial nurseryman. He served the Garden with great competence and loyalty for 23 years (1913–1936) throughout its critical formative years. A tough taskmaster, he was respected by his subordinates and admired by his superiors. Pearson the scientist and political negotiator needed just such a man as Mathews – of practical bent, with landscaping flair and strong horticultural skills. They made a perfect team, first focusing on the centrepiece, both aesthetically and ecologically, of the Garden – the Dell. The old Bird's Bath was rehabilitated. Its permanent flow was led out of the pool over a small cavern and along a graciously designed stream through the moist habitat of the Dell, which they dressed with tree ferns, flowering forest shrubs, streptocarpus, plectranthus and other shade-tolerant plants. Rough pavers, cut from the Garden's sandstone, were laid as stepping stones for visitors to traverse the cool, moist, deeply shaded retreat from summer heat. Then, as now, this is the discreet focal point of the Garden, from which all else radiates. The Garden had been given 'good bones': an elegant structure on which the natural lines and rich complexion of the plantings and landscape would mature.

Pearson's passion for gymnosperms was soon reflected in the natural amphitheatre embracing the Dell. Here his collection of over 400 cycads, donated from throughout South Africa, were planted in what remains a world-class living gene bank of these ancient and remarkable plants.

Pearson's and Mathews' sensitive understanding of the Kirstenbosch landscape – given to natural sweeps of lawns, wooded glades, flowing beds of annuals, inspiring mountain vistas and a minimum of rigid, geometric or linear arrangements – is perhaps their most fundamental contribution to the Garden's 'spirit of place'. Remarkably, for all his formal training and experience of other gardens of the day, Pearson resisted the tradition of preparing a formalised garden master plan. He trusted the talents of his team, and his successors have fortunately followed suit.

Pearson's choice of right-hand man and first Curator of Kirstenbosch was the Kew-trained Joseph 'Jimmy' Mathews, seen here in the rock garden named after him.

An early project in the Garden was the rehabilitation of the spring-fed bath built by Colonel Bird in 1811.

It might not have been so. At some point Pearson happened upon the idea of having a massive Palladian bridge (see architect's sketch on page 33) built above the Dell as a monument to Baron De Villiers – a monstrous edifice that would have dominated the Dell, and could have set a completely different course for the Garden's design. Fortunately, funds were not available for its construction, and the sanctity and tranquillity of Kirstenbosch were saved.

ABOVE *The whimsical, bird-shaped Bird's Bath remains a focal point in the Garden two centuries after its construction. The New Zealand tree fern* Dicksonia antarctica *was planted in 1914 – before a strict policy of planting only indigenous species had been adopted.*

LEFT *In this early photo, c. 1915, of the cycad collection above the Dell, note the dense poplar grove on the slopes and Silver Trees on the ridge in the background. (Courtesy of the Elliot Collection, National Archives)*

This is the place 31

Pearson's passion for cycads lives on today in the remarkable collection of fine specimens that he planted in the Cycad amphitheatre.

Brian Huntley

32   KIRSTENBOSCH

Pearson and his wife had barely settled into the beautiful home built for them on the ridge at the head of Wynberg Hill, when he died tragically on 3 November 1916 at the age of 46, exhausted from over-work, and ultimately from pneumonia following a minor operation. His peers spoke of him as

'... an enthusiastic and successful teacher, a strict disciplinarian, a man of rare charm ... It is rare indeed that men of his seriousness and earnestness of character retain so completely an almost boyish power of enjoyment that he had.'

The several photographs and portraits that remain of Pearson show a fresh-faced, boyish, gentle person with bright, intelligent eyes. He suffered no fools, and could charm and influence the powerful.

Fittingly, he was buried on a rise overlooking his beloved cycad collection, and his grave is now backed by an enormous Atlas Cedar, a gymnosperm from the mountains of Morocco, sent by his colleagues at the Royal Botanic Gardens, Kew.

Pearson's death came as a double blow to the new Botanic Garden, faced as it was with the financial and staffing difficulties occasioned by the First World War. But the Garden's strength lay not only in its excellent Director and Curator, but also in its Board of Trustees. Kirstenbosch was fortunate to have the likes of Sir Lionel Phillips, Duncan Baxter and Frank Cartwright on the first Board of Trustees, who steadied the ship with their knowledge, passion, energy and influence. In the difficult years following Pearson's death, Baxter provided the leadership, as Chair of the Trustees, and Cartwright supported the day-to-day work of the Garden. Together, they kept the Garden alive and active in the War years – until the new Director took office in March 1919.

ABOVE *Pearson died at the young age of 46. He was buried on a rise overlooking his beloved cycad collection, shaded by an Atlas Cedar* Cedrus atlantica var. glauca, *a gift sent to him from the Royal Botanic Gardens, Kew, received in Kirstenbosch and planted shortly before his death.*

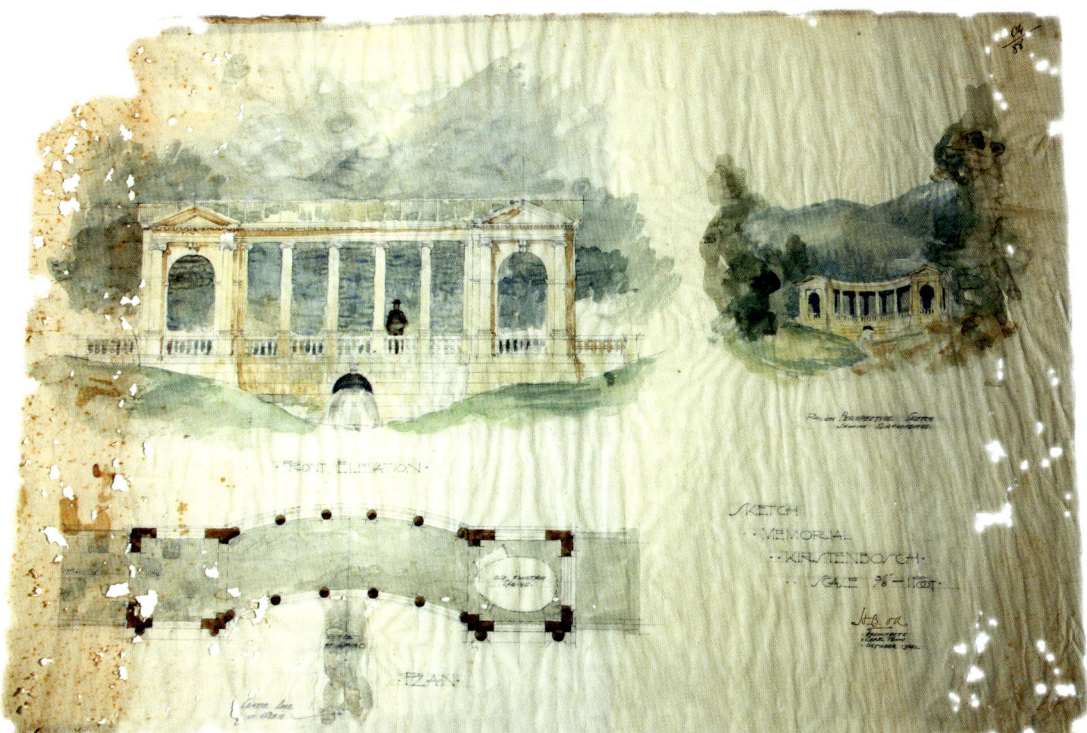

LEFT *In an apparent flight of fancy, Pearson contemplated erecting a massive Palladian bridge above the Dell as a memorial to the work of Lord De Villiers, first Chair of the Board of Trustees. Fortunately, a shortage of funds prevented implementation of this ill-conceived project. (Architect's sketch, 1914)*

This is the place 33

CHAPTER TWO

Developing the Garden

Progress from concept to reality

These three South Africans turned my thoughts to the Cape of Good Hope, not only as one of the most romantic places in the world but also a most desirable country for living in and for botanical research.

ROBERT COMPTON
A botanist's distant reminiscences – Veld & Flora, *1976*

Robust succulents, including Aloe pluridens *(in flower),* Strelitzia juncea *(spiky leaves, centre) and* Euphorbia cooperi *(back left) crowd the Mathews Rockery.*

The Compton years – no easy beginnings

Robert Harold Compton was born in Tewkesbury, England, on 6 August 1886. Like Pearson, he had a distinguished academic record at Cambridge and a special interest in gymnosperms, which took him as a young graduate to the distant Pacific – to New Caledonia. His class at Cambridge included two South Africans who later played significant roles in Cape botany – Edith Stephens and Margaret Michell (better known by her married name, Margaret Levyns). Compton was strongly influenced by these two women, and by a visit to Cambridge by Harold Pearson – as he wrote many years later (see quote on page 35).

The legacy of partnerships has been ongoing: recent collaborations with Kew – the African Plants Initiative and the Millennium Seed Bank projects, both highly successful – have been of great benefit to South African botany.

After his year in New Caledonia, Compton returned to England with a valuable cargo of botanical and other collections, most valuable of which was his young Australian bride. But he returned to a Britain at war; a committed pacifist, he joined the ambulance corps. In 1919, with his young wife and baby daughter, he set sail for the Cape, which had beckoned to him since his student years, to take up the newly created Harold Pearson Chair of Botany at the University of Cape Town, with the additional role of Director of the National Botanic Garden, Kirstenbosch. Appointed at the age of 33, he continued as Director for 34 years until 1953.

Robert Harold Compton (1886–1979)

Compton faced difficult times – the uncertainties immediately following the First World War, the Great Depression of the 1930s, the Second World War, and the normal socioeconomic and political dynamics of a developing country. Not least of his challenges was, as he later described in his 'reminiscences', his total lack of preparation for the multiple tasks at hand. As a dedicated academic, his research interests 'reflected my zeal for the pursuit of knowledge for its own sake, with a disregard for possible "practical results"'. He had no experience in money matters, staff management, or even in lecturing students. But his charm and intellect more than compensated for these initial shortcomings. Like Pearson, he was able to pursue a productive scientific programme while relying on a succession of talented curators to manage the development and day-to-day activities of the Garden.

The Marshall Ward Society at Cambridge, May 1910, included David Thoday, Edith Stephens, Robert Compton, Margaret Michell (Levyns) and R.S. Adamson, all of whom made great contributions to South African botany.

36 KIRSTENBOSCH

The Kew connection

The Royal Botanic Gardens, Kew, remain the global focus of matters botanical. South African botany has been fortunate to enjoy a close association with Kew for over 240 years.

The great patron of science in Georgian England, Joseph Banks (1743–1820), had joined Captain James Cook on the *Endeavour* from 1768 to 1771, and visited the Cape in March 1771. Banks was so impressed with the Cape flora that he recommended to King George III that a Scottish gardener, Francis Masson, be sent to the Cape to collect plants for Kew. At the time, it was the custom at Kew to employ Scottish gardeners, who were regarded as being far more thrifty and reliable than their English counterparts. This venture saw the introduction of over 400 new species into horticulture, including 50 *Pelargonium* species that form the basis of today's multimillion-dollar geranium industry in Europe.

A century after Banks' and Masson's involvement at the Cape, the Director of Kew, Sir William Turner Thiselton-Dyer, strengthened his association with southern African botany when he took on the daunting task of completing the *Flora Capensis* project initiated by William Henry Harvey in the 1860s (see page 87). Thiselton-Dyer encouraged his young assistant, Harold Pearson, to go to the Cape in 1902 to head the department of Botany at the South African College. Pearson was to drive the establishment of Kirstenbosch and, in 1913, appointed Kew-trained horticulturist Joseph William Mathews as his right-hand man. Between 1913 and 2013 four Kewites (Mathews, Thorns, Werner and Winter) served as curators, spanning a period of some 75 years.

The Palm House, the architectural showpiece of the Royal Botanic Gardens, Kew, was built between 1844 and 1848, a masterwork of the glasshouse designer Decimus Burton and Richard Turner, a Dublin iron-master.

Developing the Garden 37

During much of Compton's term, Joseph 'Jimmy' Mathews continued laying out the hard landscaping of the Garden, making full use of locally cut sandstone, including building an enormous and successful rockery; this feature was named after him in 1950, some years after his retirement in 1936. The heavy winter rainfall made the building of storm-water drains, dry-stone walls, cobbled paths and other hard landscaping essential. This was done by hand, led by Robbie Smith (master stonemason for 40 years), leaving a rich legacy to the Garden that was his abiding passion. Hot dry summers made it essential to develop reservoirs and irrigation systems, fed by the Garden's three perennial streams – its lifeblood. Mathews had started introducing and trying out indigenous plants from throughout the country – planting them in the nursery, testing their ability to grow in the particular soil and climatic conditions of Kirstenbosch, and gradually bringing them into the Garden's flower beds. The progress made in the early decades of the Garden by Mathews and his small team was, in Compton's words, 'truly phenomenal'.

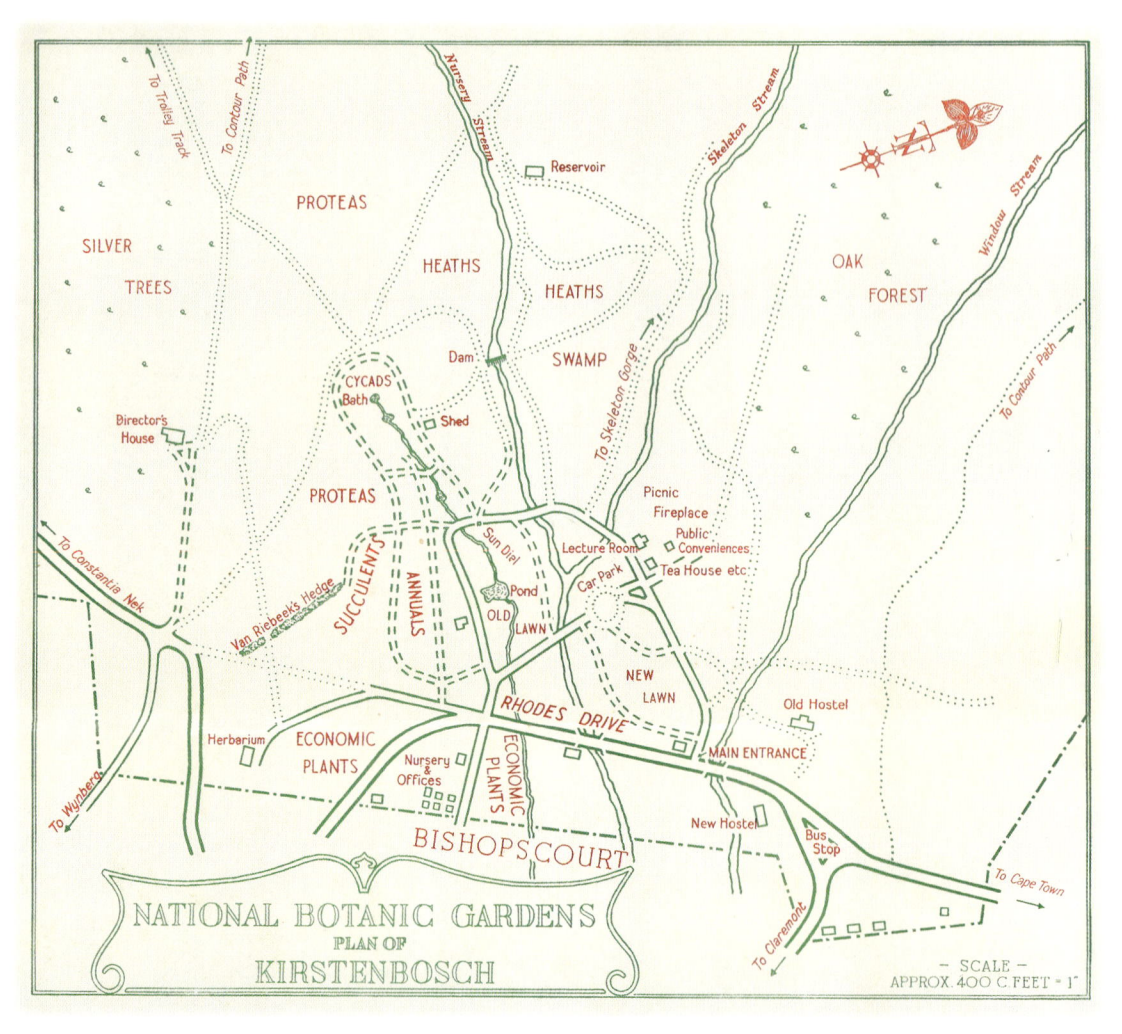

ABOVE *Robert Compton is flanked by Joseph Mathews and Duncan Baxter and members of the Board of Trustees on the occasion of Mathews' retirement in 1936. Duncan Baxter, representing the Botanical Society, served on the Board for 44 years.*

RIGHT *In this early map of the Garden (c. 1930s), note the original alignment of Rhodes Drive. Several features have disappeared since the 1930s, including the central car park, the Swamp, Oak Forest, and the Economic Plants sections, but much of the Garden's original landscape design has remained unchanged.*

38 KIRSTENBOSCH

These early views (LEFT 1930, ABOVE 1962) of one of the first landscaping projects led by Jimmy Mathews show the development of the great rockery, planted with succulents from the arid regions of South Africa, and named after its builder.

Aloe arborescens *blooms in the foreground with a large* Euphorbia ingens *in the middle distance – the same specimen as in the top photo.*

Developing the Garden   39

This wonderful painted panorama of Kirstenbosch from its earliest days appeared in the 1917 edition of the Journal of the South African Botanical Society, *forerunner of today's* Veld & Flora *magazine. Clearly indicated are: (1) the Camphor Avenue leading up towards Constantia Nek; (10) the original entrance road leading from Bishopscourt; (29) the 'Cape Chestnut Avenue', which Compton described as 'leading from nowhere to nowhere'; and (42) the old 'Trolley Track', which serviced the building of the reservoirs on the summit of Table Mountain.*

BIRDS-EYE VIEW OF THE NATIONAL BOTANIC GARDENS, KIRSTENBOSCH.

L. L. B. Goldman del.

1 Road leading to Constantia Nek
2 Road leading to Director's House and Trolley Track.
3 Road leading to Wynberg.
4 New Deviation.
5 Camphor Avenue.
6 Economic Grounds.
7 Silver Tree Forest.
8 Proposed Karroo Garden.
9 Mesembrianthemum Garden.
10 Kirstenbosch Avenue—leading to Bishop's Court.
11 Rhodes Road to Newlands.
12 Entrance Gate.
13 Bulb Garden.
14 Oak Avenue.
15 Lawn.
16 Curator's Cottage.
17 Fern Dell.
18 Nursery and Office.
19 Pearson Memorial.

40 KIRSTENBOSCH

20 Cycad Amphitheatre.	27 Ruins of Old Homestead.	34 Black Ironwood Point.	41 Nursery Gorge.
21 Protea Garden.	28 Ranger's Cottage.	35 Contour Path.	42 Trolley Track.
22 Aloe Kopje.	29 Cape Chestnut Avenue.	36 Yellow-Wood Corner.	
23 Pelargonium Garden.	30 Ruins of Col. Bird's House.	37 Stinkwood Slope.	
24 Daisy Garden.	31 Proposed Arboretum.	38 Aloe Knoll.	
25 Terrace.	32 Amaryllis Meet.	39 Window Gorge.	
26 Bolus Orchid Garden.	33 Celtis Glade.	40 Skeleton Gorge.	

The builders of Kirstenbosch

A photograph from the 1920s shows the slopes on which the protea and erica sections were developed in the 1960s.

Building the Garden was a labour of love, achieved by a loyal team of workers using limited equipment.

The builders of Kirstenbosch: in 1973 Brian Rycroft presented gold watches to eight Kirstenbosch stalwarts, each with more than 25 years' service – Abraham Basson, William Basson, Frank Krieger, David Mclean, John Fredericks, Brian Rycroft, George Basson, Nicholas Josephus and James Nicholas.

Early transport

In the early years, the only vehicle that the Garden possessed was a 'ralli-cart' – a two-wheeled driving trap for four. This carried Compton from the Garden down to Claremont Station to take the train in to the city, where a tramcar would get him to the University, then located near the present South African Museum. Mail, supplies and staff were all transported in the ralli-cart – there was no staff bus, no lorries, no tractors, no digger/loaders – just a few wheelbarrows, a sledge and a cart!

The only vehicle that the Garden possessed was a 'ralli-cart', seen here in 1932 with Winsome 'Buddy' Barker (first Curator of the Compton Herbarium), James Nicholas (trap driver) and Nelson, the loyal horse.

The Harry Bolus Bequest funded the establishment of the Herbarium built in Kirstenbosch in 1924, renamed the Compton Herbarium in 1958.

The need for accurate identification of all plant material introduced to a scientifically purposeful botanic garden required that dried specimens of each accession be preserved in the Garden's archives – in a herbarium. Developing a well-documented herbarium collection and classifying and describing the many new species being discovered became Compton's primary focus. He collaborated closely with colleagues at the University of Cape Town, where the valuable Bolus collection had been housed – under unfavourable conditions. Although it was bequeathed to the University on Bolus' death in 1911, it was decided that the collection should be transferred to a new herbarium built at Kirstenbosch in 1924. Not only did the Bolus Herbarium and Library move to Kirstenbosch, but so too did Mrs H.M.L. (Louisa) Bolus, the niece and daughter-in-law of Harry Bolus, who had appointed her Curator in perpetuity of his famous collections. Louisa Bolus made a major contribution to Kirstenbosch, not only to its science programme (see page 152), but also by initiating nature study classes (see page 153) in the Garden.

Developing the Garden

Competing ambitions in South African botany

Compton, a quietly spoken, patrician gentleman, was not a political animal. Unlike Pearson, he avoided the dynamics of the political landscape. He was probably unaware, when he arrived in the Cape in 1919, of the professional tensions between the north and the south of the country. His adversary in the politics of botany was another Cambridge man, Dr Illtyd Buller Pole Evans (1879–1968). Pole Evans, an ambitious, enormously productive and decisive red-haired Welshman, was born near Cardiff, and arrived in South Africa in 1905 to take up the post of Mycologist and Pathologist in the newly established Transvaal Department of Agriculture. Like Pearson, he started building an institution from meagre beginnings.

In 1913 the Division of Botany within the new Union government was formed under Pole Evans' leadership, and ownership of the Natal Herbarium was obtained under an agreement with John Medley Wood, the driving force for botany in that province. (The Division of Botany would evolve into the Botanical Research Institute (BRI) in 1961.) Pole Evans' ambitions were revealed when, within a month of Pearson's death, he issued a memorandum advocating the transfer of control of Kirstenbosch to Pretoria. His energy was seemingly limitless. He travelled throughout southern Africa, mapping, photographing and describing its major vegetation types and, with the backing of the Minister of Agriculture, he established the Botanical Survey Advisory Committee in 1918, with himself as Chairman. Pole Evans corresponded actively with Kew, and saw the need for strong collaboration. In 1919 he established a 'South African Botanical Liaison Officer' post at Kew, and for 73 years a succession of young South African botanists had the opportunity to work at Kew. These included many future leaders in South African taxonomy, even two future directors of the BRI – R.A. Dyer (1900–1987) and Bernard de Winter (1924–). In 1919 Pole Evans established the journal *Memoirs of the Botanical Survey of South Africa*; in 1920 he published *Flowering Plants of South Africa*, and, in 1921, *Bothalia*. He became President of the South African Association for the Advancement of Science in 1922. There could be little question that, through his decisive action and political influence, Pole Evans had developed a serious challenge to the perceived leadership of Kirstenbosch in matters botanical.

Compton was deeply traumatised by the north-south polarities that emerged when the then Prime Minister, General Jan C. Smuts, opened the new National Herbarium building in Pretoria in July 1923. The pain was exacerbated

Three champions of South African botany – Allen Dyer, I.B. Pole Evans and John Hutchinson, author of A botanist in South Africa

Dr Illtyd Buller Pole Evans, dynamic Director of the Division of Botany (later the Botanical Research Institute) in Pretoria, and one-time nemesis of Compton

by Smuts' concluding comment: 'You want a Kew. What Kew is to England and the British Empire, this National Herbarium must be to South Africa.' Worse, Smuts, a dedicated amateur botanist himself, made no reference to Kirstenbosch in his opening address. Uncharacteristically, Compton responded with some passion in an article published in the *Journal of the Botanical Society of South Africa*. Quoting at length the commentary on the matter in *Kew Bulletin*, *Nature* and the *Cape Times*, he concluded 'The Government withholds the most urgently necessary grants from Kirstenbosch, South Africa's natural Kew, while spending lavishly on the pseudo-Kew of Pretoria. It is notorious that Kirstenbosch has been and is being starved.'

Jan Smuts repaired some of the damage by his conciliatory words at the Silver Jubilee Celebrations at Kirstenbosch in 1938:

'Twenty-five years ago when the establishment of Kirstenbosch was first suggested, objections [were raised] which I need not go into to-day, and when the whole list [of objections] was exhausted the weather was used. Well, we have made a start in spite of criticisms and objections. For 25 years this garden has been in existence and it has more than justified its existence.'

Jan Smuts addresses the Botanical Society at its Silver Jubilee, celebrated in Kirstenbosch in 1938.

Developing the Garden 45

An elegant stone bell tower was erected at the gates in 1940 to commemorate the contribution made by Sir Lionel Phillips to the Garden's establishment.

In 1947, King George VI and family paid an unscheduled visit to Kirstenbosch. They were joined by Robert Compton and Jan Smuts on a visit to Maclear's Beacon, the highest point on Table Mountain, and the northeast corner of the Garden.

However, as we shall see below, the battles between north and south continued until the amalgamation of the National Botanic Gardens and the Botanical Research Institute, which created the National Botanical Institute in 1989.

Compton survived the political power play, and by the time of his retirement in 1953, Kirstenbosch enjoyed world renown. Ever humble, Compton was later to say: 'By the time of my retirement the Gardens had only, as it were, passed their adolescence.'

The first national network of botanic gardens

Hedley Brian Rycroft was the first South African Director of the National Botanic Gardens. Born in Pietermaritzburg, Natal, on 26 July 1918, Rycroft was an ecologist by training, with field research experience in the forests of Natal and in the fynbos of the Western Cape. During his 30-year tenure at Kirstenbosch (1953–1983) he oversaw the consolidation of the Garden's landscaping, and the expansion of a national network of botanical gardens across South Africa, as well as its promotion in the international botanical arena. Like his predecessors, he enjoyed the invaluable contributions of dedicated curators. Mathews had retired in 1936 and was followed by Frank Thorns (1936–1947) and H.F. Werner (1948–1959), both Kew-trained. From 1959 to 1979, Rycroft was ably assisted by Jack Marais, who in turn was succeeded by John Winter, yet another 'Kewite'.

Hedley Brian Rycroft was the first South African Director of Kirstenbosch.

46 KIRSTENBOSCH

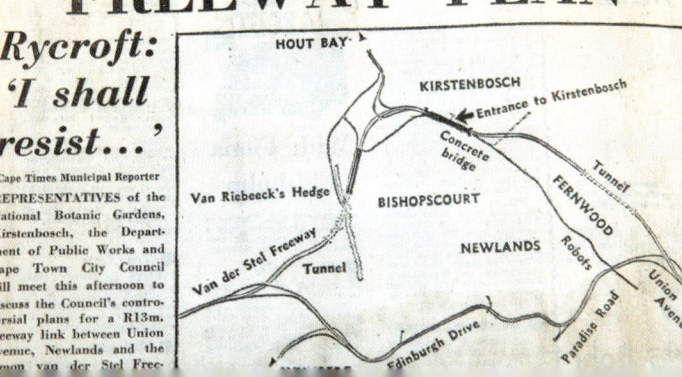

The 1950s to the 1970s saw greater government support directed towards Kirstenbosch. In 1954 the NBG had become a 'State Aided Institution' – providing more secure, but still limited, funding, and allowing many infrastructural developments to take place. By later standards, these developments tended to be relatively modest, *ad hoc* and poorly located and some required later removal or modification. But Rycroft had to grab whatever funding an unpredictable treasury made available, and these funds served the Garden well. Key improvements to the Garden included the addition of two wings to the Herbarium in 1959 (named in honour of Compton on his retirement), which received the very valuable South African Museum collection in 1956. The tradition of taxonomic research into the Cape flora was strengthened with the appointment of John Rourke to the Herbarium in 1966. He was appointed Curator from 1972 until his retirement in 2003.

Vehicle access to Kirstenbosch has been altered at regular intervals, as the demands of road usage in the area have increased. The original horse trail, along which Rhodes had planted figs, camphors, pines and chestnuts to provide shade on his excursions from Groote Schuur to Constantia, soon proved inadequate. The main road up Wynberg hill was realigned in 1914, reducing its impact on the Garden. In 1938 the entrance to the Garden at the upper end of Bishopscourt Road had to be replaced because heavy vehicular traffic was endangering pedestrian visitors. A new entrance road was constructed lower down the hill, flanked by imposing stone piers at its gate. An elegant stone bell tower was erected at the gates in 1940 to commemorate the contribution made by Sir Lionel Phillips to the Garden's establishment. A beautiful ship's bell, recovered from *HMS Dominion*, was installed, and continues to this day to announce working hours, its mellow peals being heard right across the Garden. The new access road now led up to the Tea House in the heart of the garden, to which a large car park and an unsightly toilet block were added in 1968.

The 'Battle of the Road'

One of Rycroft's greatest triumphs was preventing a freeway from being built across the Garden. The lower reaches of Kirstenbosch remained divided until 1957, when Rhodes Drive was diverted to its current position along the Garden's lower boundary. It was, at the time, assumed that the road-access problems had been finally resolved. But, in the early 1970s, the City of Cape Town initiated the development of an ambitious freeway system, one arm of which was planned to cut across Kirstenbosch. On 12 November 1971, on the occasion of the official opening of the new reservoir that had been constructed at the head of Silver Tree Ridge, Rycroft launched a vigorous campaign against the freeway's proposed routing. The media and the public of Cape Town rallied to Rycroft's call for support, and after several years the 'Battle of the Road' had been won – or so it seemed. In the mid-1980s the proposal was once again raised by the City, sending pulses of anxiety through the Garden. But by that date Kirstenbosch had been declared a National Monument, curbing any immediate prospect of the freeway's development.

Invasive aliens

When Rycroft arrived at Kirstenbosch it still had the legacy of its former use as a farm, even after 40 years of development. Exotic trees – English Oaks, Stone Pines, gums, Camphor Trees, and many aggressively invasive species competed for space – most especially on the margins of the natural forests of the Upper Kirstenbosch Nature Reserve. The Camphor Trees, Moreton Bay Figs and Stone Pines planted along Rhodes Drive remained sacrosanct, as did many of the large oaks planted by Eksteen in the 1830s. Under Rycroft's watch, a pine plantation on the southwestern boundary – Silver Tree Ridge – was harvested once mature, and the eucalypts that formed a fire belt along the ridge were gradually removed. But the real threat to the Garden was that of the invasive acacias, Cluster Pines, hakeas and brambles that infested much of its upper reaches.

Rycroft, sharing the concerns expressed by Pearson regarding invasive plants, embarked on an alien removal programme. Here the Garden's Curator, Jack Marais, played a sterling role. With a team of dedicated, energetic workmen, he implemented a vigorous clearing campaign, which Marais described as 'a magnificent obsession'. By the end of Marais' tenure in 1979, Kirstenbosch was free of the key invasive plants. In particular, his efforts led to the restoration of the indigenous forests of Nursery Ravine and Skeleton Gorge, the southernmost outliers of the great Afrotemperate Forest system that extends in a broken archipelago of forest patches from the Cape across Africa to Uganda, Ethiopia and Cameroon.

Many fine specimens of English Oaks, Stone Pines, Camphor Trees and Moreton Bay Figs gave a parkland structure to the Garden during its early years.

ABOVE *The first large cycad specimens were planted in 1914, having been railed to Cape Town from the eastern Cape at no charge by a greatly supportive South African Railways and Harbours.*

LEFT AND BELOW *Images from the Garden's formative years show the corner of the main circular path just below the Otter Pond, in 1920 and 1925 respectively.*

Unlike his two illustrious predecessors, Rycroft was not an academic. He was a pragmatist with a keen sense of the importance of positive public opinion – as demonstrated in his fight against the freeway. His passion for proteas, and indeed for all of South Africa's beautiful flowers, convinced him of the merit of sharing this heritage more broadly. Travelling widely to gardens around the world, he developed a valuable network of supporters who encouraged the display of southern African wild flowers at flower shows in their cities. Since 1976, Kirstenbosch has exhibited annually at the world's most prestigious flower show, the Royal Horticultural Society's Chelsea Flower Show in London, winning over 30 gold medals.

Following the concept of a network of botanic gardens described in Pearson's original proposal for a 'national botanic institution', Rycroft used his persuasive powers and sound knowledge of the South African vegetation and flora to expand the reach and impact of the National Botanic Gardens selectively. He was already familiar with the country's several dozen old colonial-era gardens dating from the 19th century. Few, if any, of these met the model and criteria set by Kirstenbosch. A fresh start was needed, leading on from the Karoo Botanic Garden, originally at Matjiesfontein but transferred to Worcester in 1945. Rycroft lobbied private landowners, municipal authorities and provincial and national governments for support. By the time of his retirement early in 1983, he had added five more botanical gardens to the system.

The national network would establish South Africa's NBGs as a globally unique institution: no other country could boast either a focus on its indigenous flora or a biome-based system of gardens that offered natural growing conditions, in diverse ecological settings – and for such a rich flora.

Developing the Garden

Putting science back into the Garden

In common with State-supported institutions the world over, Kirstenbosch was subject to regular government commissions and evaluations. For decades, groups of learned evaluators would study the role and performance of the organisation, make purposeful recommendations, and move on. Seldom were any of their supportive financial proposals followed – the early history of the government's relationship with Kirstenbosch was one of failed promises.

In 1976, a committee of eminent scientists was appointed to review the state of science in the National Botanic Gardens (NBG), chaired by Meiring Naudé, the former President of the Council for Scientific and Industrial Research (CSIR). The findings were harsh:

> 'Research does not actually exist at the National Botanic Gardens of South Africa ... research is only being undertaken in the area of taxonomy, and even here it is totally inadequate, especially in the light of the abundance of available material.'

The situation was, in fairness, more a reflection of the extremely limited budget available to the NBG compared, for example, with its northern rival, the Botanical Research Institute (BRI). In 1976, the NBG had a single herbarium of 200 000 specimens, and a single taxonomist. The BRI, in contrast, had the National Herbarium in Pretoria, plus regional herbaria in Durban and Stellenbosch, and a management agreement with the herbaria of Grahamstown, Kimberley and Windhoek, with collections of over one million specimens and more than 30 researchers – taxonomists, ecologists, anatomists and palaeobotanists.

The Board of the NBG was somewhat sobered by the Meiring Naudé Report and, in selecting a successor to Rycroft, decided that an academic with a broader research background than that of taxonomy would be appropriate. Professor J.N. (Kobus) Eloff took up the directorship vacated by Rycroft on his retirement early in 1983.

Kobus Eloff, born in Johannesburg in 1939, joined Kirstenbosch on 1 August 1983. He came from the University of the Orange Free State, where, as Head of the Department of Botany, he had conducted an active programme in plant physiological and biochemical research, with a special interest in medicinal and toxic plants. Recognising the particular challenges and needs of the NBG, he set out a plan of action with four principal strategic areas – research, plant utilisation, horticulture and education. Eloff's arrival was well timed: shortly before he joined NBG, the organisation had passed from the political limbo of a state-aided institution located within a variety of ministries, to the jurisdiction of the Department of Environmental Affairs and Tourism, where, after several incarnations, it remains to this day. By a peculiar rotation of ministerial responsibilities, the legal persona of the NBG had been promulgated not within the Ministry of Environmental Affairs, but through the Forest Act No. 122 of 1984 – a situation that has led to some confusion over the years. But from 1984, as a statutory Board, the organisation had a more secure future in the country's changing landscape of science politics.

A key challenge to mobilising a research programme at Kirstenbosch was the lack of staff and facilities. Eloff relieved

Kobus Eloff, Director from 1983 to 1989, promoted a more scientific approach to proceedings in the NBG.

A new generation of Kirstenbosch research was initiated under Kobus Eloff at the Threatened Plants Research Laboratory.

A succession of planning exercises

In 1973, during Rycroft's tenure, the Cape Town architectural firm of Gabriel Fagan undertook a comprehensive and detailed study of the Garden, its development history, physical attributes, socioeconomic location and linkages. It recommended that most of the developments should be located between the Moreton Bay Fig and Camphor Avenue and the new Rhodes Drive. A new 5 500-square-metre National Botanic Centre would be placed within the garden, on the north-facing slope above the original Garden entrance gates. The Fagan plan recommendations were, in turn, referred to an architectural firm that was commissioned to draw up sketch plans for the proposed buildings. The sketch plans revealed that the developments would result in too great an impact on the Garden, and so another development plan was commissioned – this time from the Pretoria-based landscape architect Roelf Botha. The Botha plans were presented to the Board in early 1981, with the location of the proposed developments north of Window stream, near Pearson House.

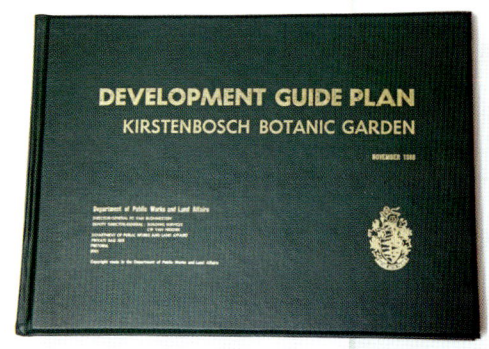

A seemingly endless series of meetings of the Board, building subcommittees, advisory committees, liaison committees and the Botanical Society Council studied various options of location, design and materials – without reaching agreement. The proposed location of key buildings hopped from within the Garden, to land across Rhodes Drive to the northeast of the Stone Cottages, to a disputed area of land known as 'Newlands Heights' – and back to the Garden. A controversial issue was that the government funding agency – the Department of Community Development – insisted that all buildings should be thatched – in harmony with what the planners in Pretoria perceived as the 'Cape architectural style'. The Board had great difficulty in convincing the Pretoria officialdom that to thatch a new herbarium would put its irreplaceable collections at risk. A few months later, when the thatched Tea House burnt to the ground in 1982, their point was adequately made.

There was no shortage of planning exercises.

Developing the Garden

Protea Village forced removals

While apartheid will forever be the dark stain over South Africa's history, its shadow fell heavily on Kirstenbosch in 1967 with the forced removal of many families of loyal Kirstenbosch staff from Protea Village, a small community living in the land below the Garden's eastern boundary. The government's policy of racially segregated 'Group Areas' was enforced across the country, and the Protea Village community was transferred to the Cape Flats. Kirstenbosch was fortunate to retain the profound loyalty of its long-serving staff. Despite the trauma of dislocation from the peace and convenience of living on the land adjoining the Garden, the people of Protea Village continued to serve the Garden with devotion. Justice was done in 2004, through the restitution of land rights to 80 Protea Village families, and the opportunity to reclaim title to land that had been home to many families for over a century.

THE VANISHING COTTAGES

THIS IS ONE OF THE COTTAGES on Kirstenbosch Drive which will disappear during the next few months. Mrs. Katie Smith (60) was born here and has lived here ever since.

Cape Argus, 27 May 1967

the staffing situation by attracting young graduates to short-term research projects that brought early results. He also raised funds from the Rowland and Leta Hill Trust that enabled the construction of a Threatened Plants Research Laboratory – the incubator of a wider research agenda. But much more was needed to transform Kirstenbosch from a beautiful garden, with interesting plants, splendid vistas and a herbarium, into a Botanical Garden for the 21st century. For this, serious money, invested in a grand plan, was needed. Eloff, with Board support, focused on two initiatives – infrastructure development and plant production.

Eloff was soon to learn that Compton's succinct view of the story of Kirstenbosch as one of 'romance and hard facts' could be paraphrased as one of 'realities and hard lessons'. Hardest of these was the failure of a succession of development planning exercises commissioned by the NBG Board between 1973 and 1988.

The seemingly endless debate around development plans continued to the end of Rycroft's watch in 1983. Eloff, however, had an expanded vision for the organisation, with new requirements for increased research and education facilities that changed the field of play. In 1985, yet another firm of landscape architects, Chris Mulder Associates, also from Pretoria, was commissioned to review the Fagan and Botha proposals (see box on page 51) and, in consultation with the new Director, presented a new master plan. During ongoing discussions with all parties, it emerged that the Department of Environmental Affairs and Tourism had approved a budget of R50 230 000 for the project. The euphoria was short-lived. Further enquiries revealed that the figure was a figment of some Pretoria official's imagination. Needless to say, the funding never materialised. Submitted in 1988, Mulder's proposals constituted a massive tome with ambitious ideas of relocating virtually all buildings within the Garden. Further, many new facilities, as well as staff accommodation, would be located outside the Garden on land from which the Protea Village community (including many of the Garden's loyal staff) had been forcibly evicted in the mid-1960s (see page 52). But by the end of the 1980s, the ambitious Kirstenbosch Development Plan had been abandoned, at least until a new direction could be found.

Compton had learned after two decades of hard work that 'economic botany' had little chance of success, given the small budget available; and, more especially, that the proper place for such work would be under the auspices of the Department of Agriculture. Eloff took a somewhat different approach, and championed the development of the commercial potential in horticulture of South Africa's indigenous plants. The 'Plant Production' division hoped to use modern tissue-culture technologies to rapidly propagate new selections of indigenous plants, while also focusing on securing the future of threatened species through a programme of 'conservation through cultivation'. A seed bank for rare plants was initiated, and students and facilities were funded by special grants and donations. But the exercise was hopelessly undercapitalised, and resistance from the commercial horticulture sector presented Eloff with an ongoing challenge. Despite some remarkable successes – such as the discovery by researcher Hannes de Lange of the smoke-based germination cue for fynbos species (see page 176) – the programme failed to produce the desired results.

While the NBG Board, and especially Eloff and his colleagues at Kirstenbosch, must have found the 1980s particularly frustrating, their hard work laid the foundations for later successes.

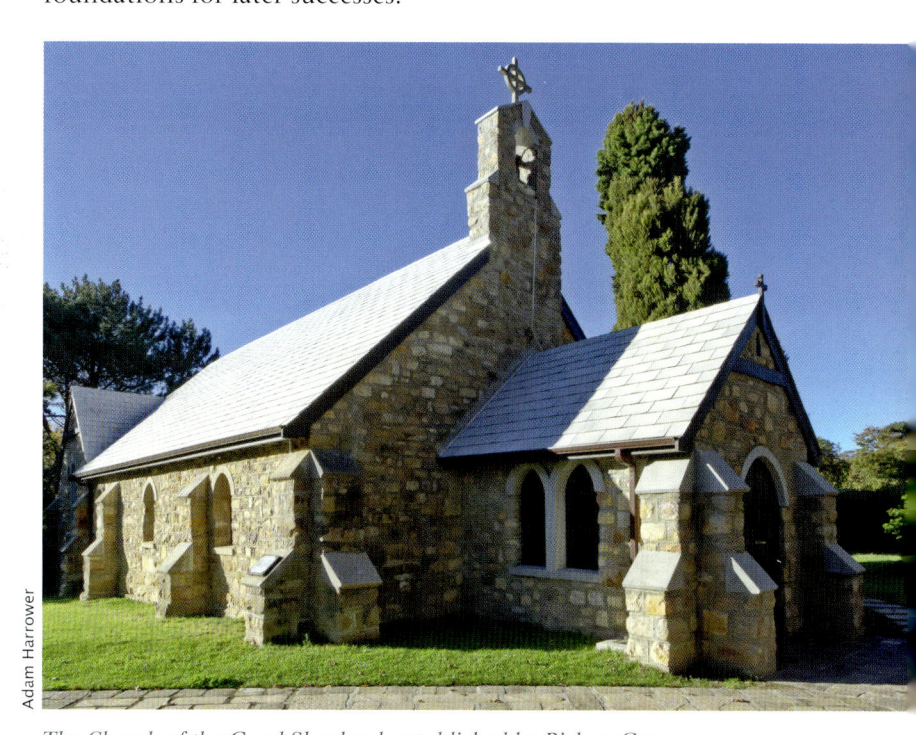

The Church of the Good Shepherd, established by Bishop Gray in 1867, is still a spiritual anchor for many former Protea Village residents who gather annually for 'Kirstenbosch Sunday' on the first Sunday in spring.

Developing the Garden

From National Botanic Gardens to National Botanical Institute

Parallel to the process of planning the infrastructural development of the Garden, a far more challenging process was being advanced by the NBG team under the leadership of Eloff. The continuing problems of financing the institution no doubt led the policymakers within government to question the NBG's relationship with other players in the botanical arena. By 1985, the BRI had established a foothold in Cape Town, with its then Experimental Ecology Group being housed at the University of Cape Town, while within the Department of Agriculture as a whole, the various research institutes were being incorporated into the new Agricultural Research Council. The BRI was 'offered' to the NBG. An overall review of the NBG's needs in 1987 was expanded to consider how botanical research, conservation, education and plant commercialisation activities in South Africa could best be co-ordinated and improved.

The result was the most dramatic change in the history of the country's institutional arrangements for botany.

Using the legal framework provided by the Forest Act of 1984, the Department of Environment Affairs and Tourism was able to bring about the amalgamation, in 1989, of the National Botanic Gardens and the Botanical Research Institute. The two institutions had been polarised and overtly competitive, if not actually antagonistic, for 75 years. Following lengthy negotiations between the two key government departments – Environment and Agriculture – the personnel and assets of the much larger BRI were incorporated within those of the modest NBG to form the National Botanical Institute from September 1989. Contrary to many expectations of a Pretoria base for the new organisation, the decision was made to place its head office in Kirstenbosch.

Three quarters of a century after initiating the development of a truly national, truly botanical, and truly indigenous South African institution, Harold Pearson's dream had been realised. Before considering developments in the last decades of Kirstenbosch's first century, let us explore the botanical riches that have made the Cape Floral Kingdom so famous.

Plants from diverse habitats, such as the desert-loving sparkling mesem Drosanthemum bicolor *and the delicate, wetlands* Crinum campanulatum, *flourish in the Garden.*

54 KIRSTENBOSCH

Mass displays of protea species, such as the popular Leucospermum cordifolium, *are among the most photographed plants in the Garden.*

Adam Harrower

Developing the Garden   55

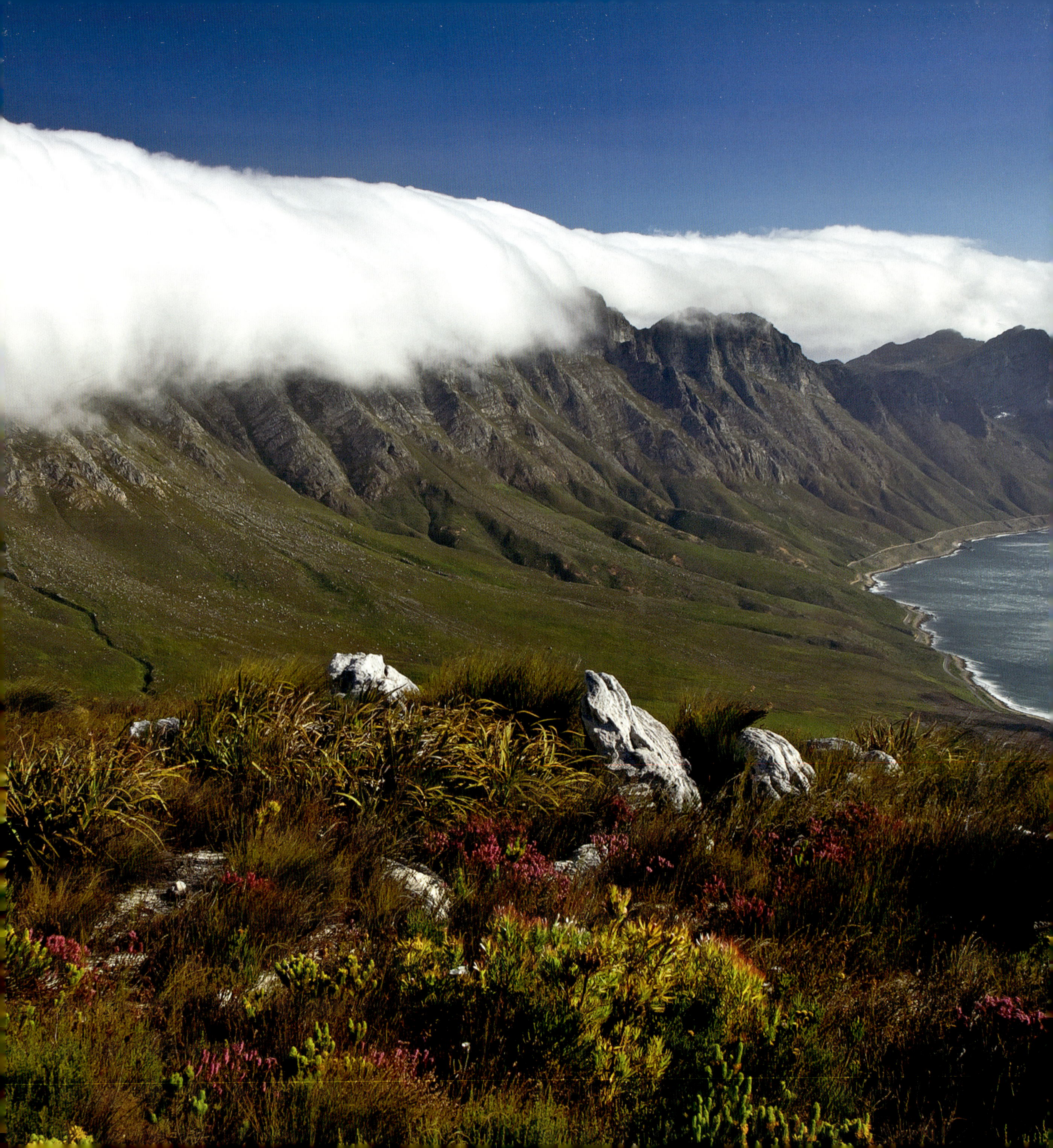

CHAPTER THREE

The Cape Floral Kingdom

The world's hottest 'hot spot' of botanical diversity

... this land of the Cape of Good Hope in farthest Africa no botanist ever before had trod. Oh Lord, how many, how rare and how wonderful were the plants that presented themselves to Hermann's eyes! In a few days Hermann simply and solely discovered more new African plants than all the botanists who ever before him made their appearance in the world.

CAROLUS LINNAEUS

Linnaeus' description in his Flora Zeylandica, *1747, of the discoveries of Paul Hermann, the first professional botanist to visit the Cape, which he did while en route to Ceylon (Sri Lanka) in 1672*

The Kogelberg, at the epicentre of the Cape Floral Kingdom, has an extremely rich diversity of ericas, restios and proteas. In midsummer, southeasterlies off the Agulhas Current blanket the Kogelberg in cloud.

The Cape Floral Kingdom

South Africa has many icons – outstanding natural features such as Table Mountain and the Drakensberg, tourist destinations such as the Kruger National Park and Kirstenbosch, sports teams (the Springboks and Bafana Bafana), and heroes of the calibre of Nelson Mandela and Desmond Tutu. All of these provide positive brand names and images of the country. In terms of our biodiversity, there is no more potent icon than that of the Cape Floral Kingdom.

Today, some scientists prefer the designation 'Cape Floristic Region' but we will use, in this chapter at least, the more romantic 'Cape Floral Kingdom'. Perhaps because the term was first published as *Reich der Kapflora* in 1908 in an obscure, inaccessible German report – the *Wissenschaftliche Ergebnisse der deutschen Tiefsee-Expedition auf dem Dampfer Valdivia 1898–1899*, in the university town of Jena, its origin has been overlooked by most South African scientists. The report's author, Rudolf Marloth, has been described by many as the 'father of modern botany in South Africa'.

Rudolf Marloth (1855–1931), botanist, chemist, ecologist and biogeographer

Floral kingdoms of the world

A floral kingdom is defined by the species, genera and families that occur within its boundaries, and nowhere else – i.e. it is defined by the composition of its endemic flora, not by the growth forms and structure of its vegetation.

Six floral kingdoms are recognised – Boreal or Holarctic (Eurasia and North America); Palaeotropic (Africa); Neotropic (South and Central America); Australasian; Antarctic; and the Cape Floral Kingdom (CFK). The Boreal Kingdom covers over 42 per cent of the world's land surface area, and the CFK covers less than 0.04 per cent – a minute crescent of land along the extreme southwest and south coast of the African continent. But while the Boreal

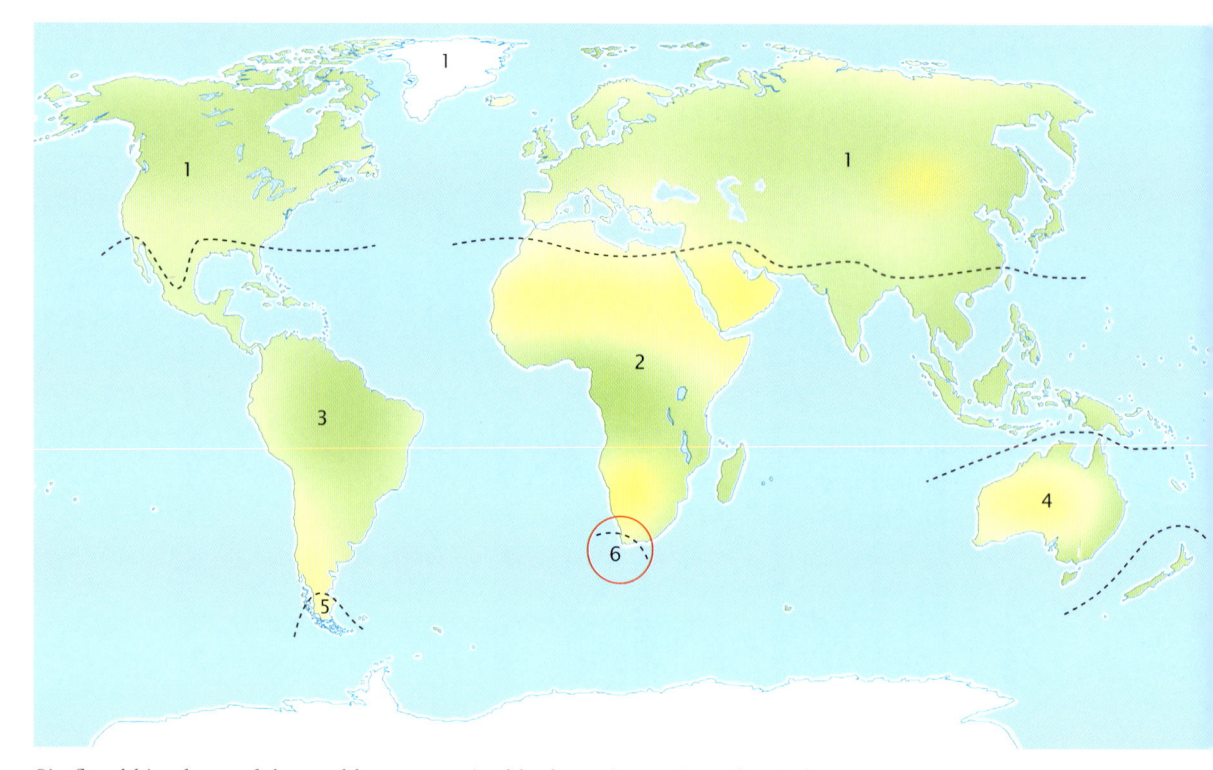

Six floral kingdoms of the world are recognised by botanists: 1 Boreal; 2 Palaeotropic; 3 Neotropic; 4 Australasian; 5 Antarctic; 6 Cape. The Cape Floral Kingdom is a tiny speck at the southern tip of Africa, and the world's tiniest plant kingdom.

Rudolf Marloth was born in Lübben, Germany, in 1855. He studied pharmacology in Germany and Switzerland, but poor health persuaded him to travel to warmer climes. On 30 December 1883 he arrived in Cape Town. Two days after the long sea voyage, he climbed Table Mountain, and started assembling a herbarium that was to expand to 15 000 specimens before his death in 1931. An indefatigable mountaineer and an acutely observant natural historian, he soon knew the Cape, and much of southern Africa, better than the locals. He had worked with the great plant geographer Andreas Franz Schimper, and had undoubtedly read much of naturalist Alexander von Humbolt's pioneering work on vegetation patterns, altitudinal zonation and climatic and other determinants of plant distribution – sufficient for him to recognise the globally unique character of the Cape flora, and coin the term 'Cape Floral Kingdom'.

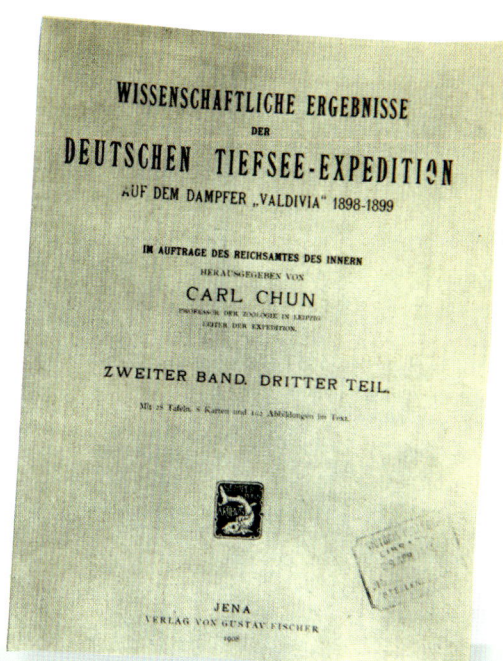

Marloth, in his classic contribution to the scientific reports published in 1908 of the German 'Valdivia' expedition, was the first botanist to define and describe the concept of a Cape Floral Kingdom.

Kingdom has a flora of some 45 000 species, that of the CFK has 9 381 – roughly 16 times the species density of the Boreal. More impressive are the high levels of endemism in the CFK, with over 68 per cent of its flora found nowhere else on the planet, and comprising some 154 endemic genera (whole genera that occur nowhere else on Earth) and six endemic or near-endemic families (again, plant families that are found nowhere else on Earth). But this flora is under threat, with no fewer than 1 736 of its species being assessed in danger of extinction if current trends of land transformation continue.

The Palmiet River drains the Kogelberg Biosphere Reserve in the heartland of the Cape Floral Kingdom, where 1 882 species of flowering plant and fern are found within less than 900 square kilometres.

The Cape Floral Kingdom 59

The Cape Floral Kingdom is 90 760 square kilometres in extent. It stretches in a narrow band from near Vanrhynsdorp in the north, south to the Cape Peninsula, and eastwards through the southern Cape as far as Grahamstown. It lies within 200 kilometres of the coast, following the Cape Fold Mountains of the Western Cape. The CFK is dominated by the Fynbos Biome (82 per cent), with patches of Succulent Karoo (12 per cent), Subtropical Albany Thicket (3 per cent) and Afrotemperate Forest (0.1 per cent).

The distinction between the Cape Floral Kingdom and its dominant vegetation – fynbos – is subtle and often confused, or confusing. Although the CFK is defined largely by the presence of fynbos, its boundaries also encompass a much smaller component of other vegetation types, as detailed above. Scattered islands of fynbos also occur beyond the CFK. Thus, although fynbos and the CFK are almost, but not quite, synonymous, the now more commonly used term for the region is 'fynbos'.

The Fynbos Biome

In the 1970s, a suite of detailed interdisciplinary and collaborative research projects was co-ordinated by the CSIR. These became known as the Ecosystem or, later, the Biome Projects. First was the Savanna Ecosystem Project at Nylsvley; the Fynbos Biome Project was next, followed by similar projects investigating the Karoo, Forest and Grassland biomes. Their impact on South African biodiversity science has been profound.

The leading botanists and zoologists of the time argued endlessly on the choice of name for the biome. Terms such as capensis, maquis, sclerophyllous bush, heathland or even macchia (then in use in the standard text, John Acocks' classic *Veld Types of South Africa*) were rejected. Fred Kruger, then Director of the South African Forestry Research Institute at Jonkershoek, championed the use of the traditional term 'fynbos' (from the Dutch *fijnbos*, pronounced 'fayne-boss') – and won the day. Ironically, the term was originally somewhat scathingly applied by early settlers to any sort of woodland lacking in timber trees, a critical resource; but it soon replaced its synonyms in all scientific and popular literature. The enthusiasm for the Cape flora and vegetation was reflected in the success of the Fynbos Biome Project, annual meetings of which continue to this day as the Fynbos Forum.

The term 'biome' was also introduced into South African scientific literature in the 1970s. A biome is the highest category of vegetation recognised at continental scale, defined according to the dominant growth form and structure of its vegetation and the major climatic features that affect this. It focuses on structure and function, not floristics, which is concerned with the distribution of individual species.

Diverse fynbos abounds in the Kogelberg Biosphere Reserve, inscribed by UNESCO in 1998 as the first Biosphere Reserve in South Africa.

Amida Johns

The Cape Floral Kingdom

The vegetation map of South Africa, Lesotho and Swaziland

In one of the most important contributions to the region's biodiversity literature of the new millennium, researchers at the Kirstenbosch Research Centre, in collaboration with over 100 ecologists across the country, completed a 10-year project to produce a monumental encyclopaedia – *The Vegetation of South Africa, Lesotho and Swaziland*. Published in 2006 under the leadership of Michael Rutherford at the Kirstenbosch Research Centre and Ladislav Mucina at the University of Stellenbosch, the massive volume describes the ecology, composition, conservation status and distribution of the nine biomes and 435 vegetation types defined in the survey area, with a detailed map published at a scale of 1:1 million. The study defines nine biomes for southern Africa – Desert, Succulent Karoo, Nama-Karoo, Fynbos, Grassland, Savanna, Albany Thicket, Indian Ocean Coastal Belt and Forest. It is without doubt the finest treatise on any regional vegetation today.

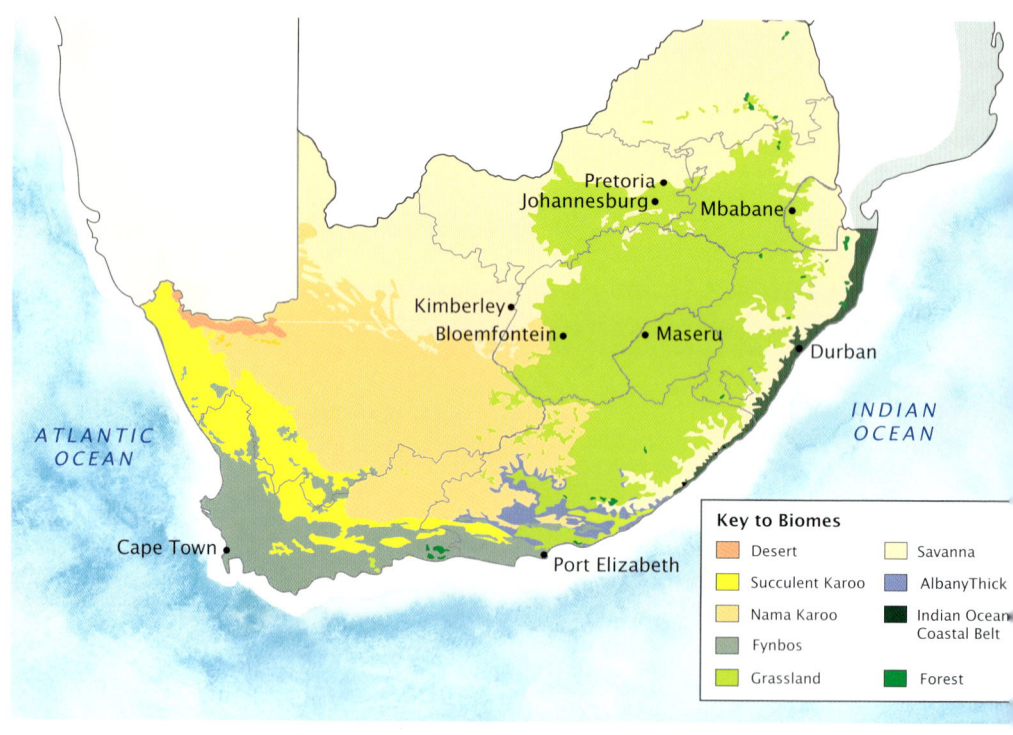

The Vegetation of South Africa, Lesotho and Swaziland – *a monumental treatise*

The nine biomes of South Africa

Desert

Succulent Karoo

62 KIRSTENBOSCH

Nama-Karoo

Fynbos

Grassland

Savanna

Albany Thicket

Indian Ocean Coastal Belt

Forest

The Cape Floral Kingdom 63

The surge of interest in, and research on, the ecology of fynbos stimulated by the Fynbos Biome Project produced a new generation of scientists – many of these now in leading positions in the country's academic and research communities. Few have contributed more to our understanding of biomes than Richard Cowling, a young B.Sc Honours student at the time of the initiation of the project, and now globally recognised as the foremost authority on the subject. Cowling and his colleagues, William Bond, David Richardson and many others at the University of Cape Town during the exciting years of the 1980s and 1990s, have condensed the huge body of existing and new research into many papers in scientific journals, volumes of scientific syntheses and popular literature. Yet nowhere is there a short, simple, succinct answer to questions about the richness and the relative treelessness of the fynbos.

Three basic ecological factors determine plant life and ecosystem dynamics in fynbos. First is climate, characterised by hot dry summers with strong southeasterly winds, and cool wet winters, with gale-force northwesterlies. The extent of summer drought is variable, most intense in the west, decreasing eastwards across mountains but always a key determinant for plant growth and ecosystem dynamics. Second is the soil, which comprises shallow sandstone and granitic soils, low in essential nutrients. The soils, too, are variable, following the underlying geology, but for the most part are far poorer than the soils of other Mediterranean-type ecosystems – with the exception of those of southwestern Australia. And third is the regular occurrence of fires that are extensive and of high intensity. Fire, as in all Mediterranean ecosystems, is a temporally and spatially variable factor, but is key to the very existence of fynbos.

The amazing species diversity of the Fynbos Biome

For centuries, botanists have wrestled with the paradox of a species-rich, treeless flora on the windswept, nutrient-deficient soils of the Cape. But the southwestern Cape, home to the Cape Floral Kingdom and to the Fynbos Biome, was not always a place of winter rains and open, treeless heathlands.

To understand the origins of the Cape Floral Kingdom and fynbos one must go back at least to the great events of 65 million years ago, when the age of the dinosaurs and gymnosperms came to a calamitous end, and the age of flowering plants and mammals took over.

For much of the last 65 million years, moist tropical forests dominated the landscapes of what is now southern Africa. Many familiar genera such as *Podocarpus, Protea, Araucaria, Casuarina* and tropical families like Anacardiaceae, Caesalpiniaceae and Euphorbiaceae, plus many species of palm, are present in the fossil record of this period. The climate oscillated through these many millions of years, with the expansion and contraction of closed forests and more open vegetation formations. Big changes started to happen from about 16 million years ago, with the final separation of Antarctica from South America resulting in the Antarctic Circumpolar Current, and later, about 12 million years ago, the cold Benguela Current.

Even more significant were the northward migration of the African continent, the build-up of the Antarctic icecap, and the lowering, and subsequent rising, of the sea level – creating extensive coastal plains of sandy, leached soils and, in some places such as the Agulhas Plain,

Gladiolus virgatus *is a stunningly beautiful gladiolus found on a few mountains of the Helderberg and Du Toit's Kloof – first known only from material bought in 1907 from flower sellers in Cape Town, and not described until 1993.*

64   KIRSTENBOSCH

Fynbos families

Floristically, fynbos is characterised by three families – Restionaceae, Proteaceae and Ericaceae, plus a vast array of species from other families, notably Iridaceae, that share a common strategy to survive the summer drought by having underground bulbs, corms or rhizomes – the so-called geophytes.

Restios – also called Cape reeds – are grass-like but with much reduced leaves, and are found in all fynbos communities, from the coastal shores to mountain peaks. Restios have been around for over 120 million years, and are also found in Australia. The fynbos is home to 314 species of restio – all with separate male and female plants.

A major contributor to the diversity of fynbos is that of the genus *Erica* – with 682 out of a global total of 816 species. No fewer than 104 species are to be found on the Cape Peninsula. Europe has just 14 species, of which only six occur in the much romanticised heathlands of Scotland. For several months of the year, ericas look drab, grey-green and monotonous, with very little diversity in growth form and dimension. But from spring to the end of autumn, at flowering, the swathes of colour over the mountains and plains – pinks and mauves and complex mixes of greens and yellows – give a special serenity and tranquillity to the scene.

Members of the Proteaceae offer the closest approximation of trees in fynbos. Some Leucadendrons, such as the famously beautiful Silver Tree, *Leucadendron argenteum*, attain 10 metres in height on the mid-slopes of Kirstenbosch. Restricted to granite outcrops on the Cape Peninsula, this lovely species was extensively felled for fuel well into the 19th century, and is now much reduced in abundance and occurrence. The protea family has 387 species in South Africa, of which 90 per cent occur in fynbos.

The geophytes are perhaps the most spectacularly beautiful of the Cape's treasure trove of floral wonders. With some 2 200 species in the Cape, they have contributed greatly to the glamour and colour of gardens and homes in Europe – species, cultivars and hybrids of many members of gladiolus, watsonia, freesia, lachenalia, ornithogalum, ixia, amaryllis, etc, having been introduced abroad since the late 1700s.

The four major groups within the fynbos are represented here by (FROM TOP TO BOTTOM): Cannomois grandis *(Restionaceae),* Leucospermum erubescens *(Proteaceae),* Erica regia *(Ericaceae) and* Sparaxis elegans *(geophytes).*

Fire in fynbos is as ancient as the biome itself. Kirstenbosch has had frequent fires on the upper slopes of the Garden; here two helicopters in close formation help control the flames by dropping water from above.

Alice Notten

66 KIRSTENBOSCH

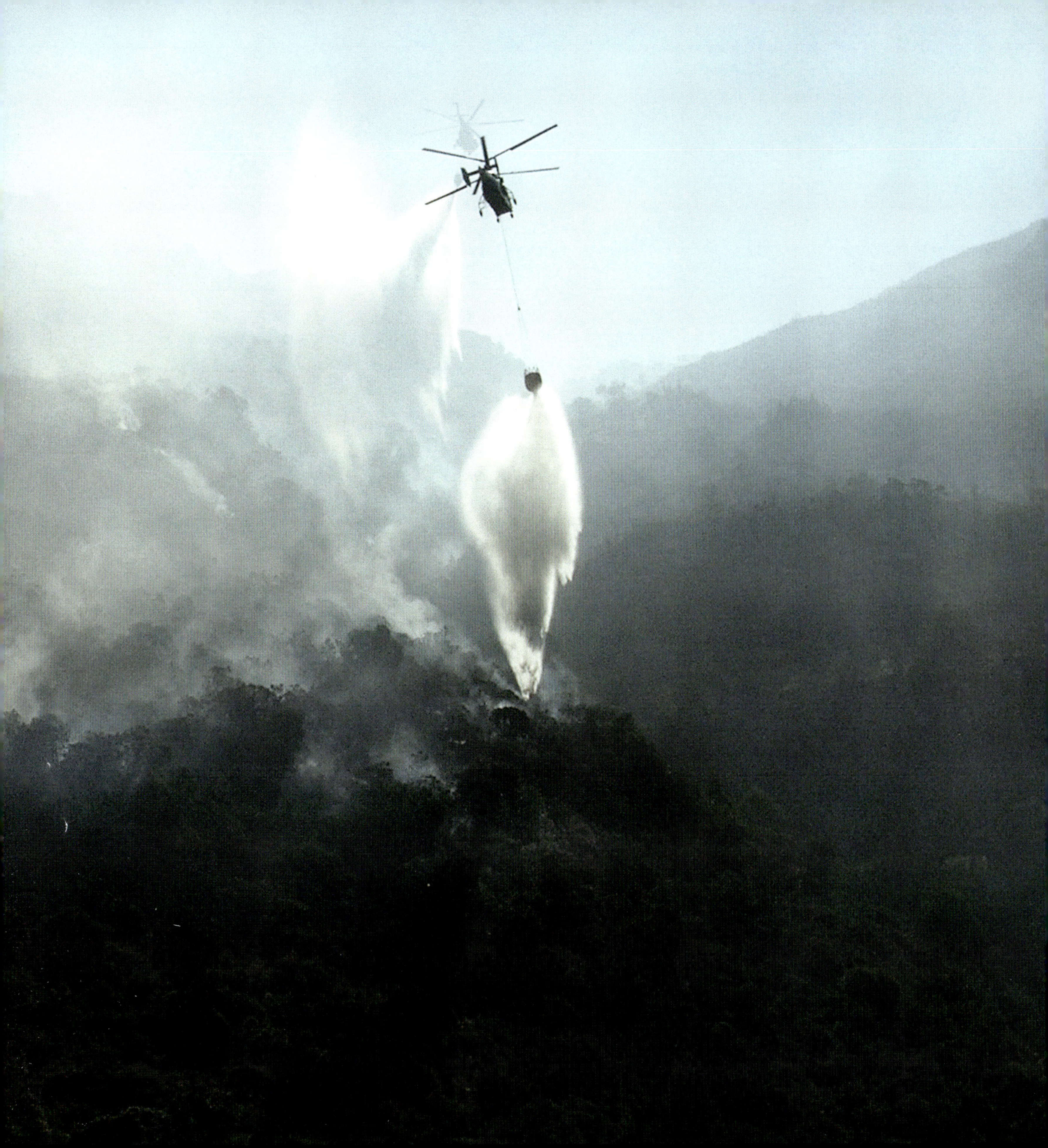

limestone. By five million years ago, drier and generally cooler conditions prevailed, with cool, wet winters, and warm, dry summers. Fires became a significant driving force in ecosystem dynamics, opening up the vegetation and introducing clear habitats into which new, rapidly speciating families such as grasses and daisies, as well as shrubs and geophytes, could compete with fire-sensitive forest species. As the role of fire increased, soil erosion intensified, and soils became impoverished, again providing new opportunities for plant diversification.

Fingerprints of the past are to be found in what are known as palaeo-endemics. In the Cape, the endemic families Geissolomaceae, Grubbiaceae and Roridulaceae, and the near-endemic Bruniaceae and Lanariaceae, represent ancient groups from among the earliest flowering plants. Unlike the old flora of the Boreal Kingdom of Eurasia and North America, which was largely eliminated by the ice sheets and glaciers of the Pleistocene Ice Age – the African flora escaped the long glacial events of the Pleistocene. When the temperatures dropped by 5°C over the 100 000-year-long glacials, cold-sensitive species were able to retreat down mountainsides, and ascend again as the climate ameliorated during the warmer, but much shorter, 10 000-year interglacials. It was this warming and cooling during the Pleistocene, over a period of some 1.5 million years, that accelerated speciation in the Cape flora.

Other characteristics of the region came into play: the diversity of landscapes, with mountain ranges rising to 2 000 metres close to the coast; a geological mix of coarse sandstones and soft shales; acidic and nutrient-poor soils, moderately fertile clays, and calcareous limestones; a variable climate, cool with high rainfall in midwinter, to arid, hot interior valleys in summer; and a mix of fire regimes. All provided the opportunity for rapid radiation of the flora on the interface between the older summer-rainfall areas of southern Africa and the emerging winter-rainfall zone that has developed over the past five million years. Within this area, many new endemics (or neo-endemics) within the families Mesembryanthemaceae, Ericaceae, Iridaceae, Proteaceae, Rutaceae, Rhamnaceae, Fabaceae, Polygalaceae and Rosaceae display the results of this rapid and recent speciation. Many of these species occur in sites of very limited area, often on mountain tops – with many peaks having their own individual endemics – or on the localised limestone outcrops of the Agulhas Plain.

Cowling explains the rapid speciation of short-lived fynbos shrubs by referring to the isolation of gene pools of populations subject to recurrent fire events, during which a whole generation of a species might be killed. Species that depend on ants to disperse and bury their seeds re-establish as isolated patches, thus avoiding both intergenerational gene flow (because no mature plants survive the fire) and the introduction of genes from distant populations, because of the limited area occupied by individual populations. Such 'catastrophic' evolutionary events, where populations go through a genetic bottleneck, are conducive to the formation of new gene complexes, leading to the evolution of new species. The steep environmental gradients found in the Fynbos Biome offer multiple opportunities for such isolation events.

A narrow endemic, the famously beautiful Silver Tree Leucadendron argenteum *grows naturally only on the Cape Peninsula.*

The African origins of fynbos

Contrary to earlier opinion, fynbos flora did not radiate from an ancient 'southern' flora, but rather from the families that were already well established in Africa. Ecological changes opened up new habitats, and evolution did the rest. There are several 'Gondwana' taxa in the Cape flora – such as *Podocarpus* and *Widdringtonia*, and South Africa shares several families, such as Proteaceae and Restionaceae with Australia, but these families evolved after the drifting apart of Gondwana, and probably indicate dispersal across narrow seas between the continents at an early stage of the break-up.

Retzia capensis *(in the endemic family Stilbaceae) represents an ancient group of angiosperms.*

Brunia stokoei *is a member of the family Bruniaceae, and is an endemic of the Kogelberg in the Western Cape.*

The hottest hot spot of them all

In 1992, at Rio de Janeiro, Brazil, world leaders gathered to define a new strategy to save the diversity of life on planet Earth. The United Nations Conference on Environment and Development (UNCED) approved a suite of major global programmes to address problems of biodiversity loss, desertification and climate change. The outcomes of the meeting – popularly know as the 'Earth Summit' – provided South Africa with a unique opportunity to align its new national policies and legislation in the post-1994 law reform programme. As a result, South Africa has, today, one of the most advanced, comprehensive, and sophisticated series of environmental laws to be found in any country.

A key agreement signed at the 1992 UNCED is the Convention on Biological Diversity (CBD). The Convention defines biodiversity as 'the variability among living organisms and the ecological complexes of which they are part; this includes diversity within species, between species and of ecosystems'. The Convention's objectives are 'the conservation of biological diversity, the sustainable use of its components and the fair and equitable sharing of the benefits arising out of the utilisation of genetic resources'. It is these concepts that underpin global, regional and national conservation strategies around the globe today.

The Cape Floral Kingdom

The Cape Sugarbird is endemic to the Cape Floral Kingdom.

Important to South Africa was the financial mechanism established at Rio to assist developing countries to meet the objectives of the CBD. Known as the Global Environment Facility (GEF), it has contributed, since 1996, over $150 million to biodiversity conservation projects in South Africa – the majority of these within the Cape Floral Kingdom.

But even with GEF resources, the challenges of preventing the extinction of the world's diversity of life will require not millions, but billions of dollars. Focus is needed: 'the problem is very big, the fuse is very short', in the words of Tom Lovejoy, Amazon scientist, conservationist and a champion of the fynbos. It was another fynbos admirer, and regular visitor to Kirstenbosch, Norman Myers, a British environmentalist specialising in biodiversity, who came up with an interesting proposal to speed up the rescue of the 'sinking ark'.

In a paper published in 1988, Myers identified 10 global 'hot spots' – sites within the most threatened, but biodiverse, tropical rainforests – where the biggest 'bang for the buck' could be achieved in terms of species protected from extinction. The concept of hot spots was immediately embraced by a new organisation, Conservation International, led by Russell Mittermeier, another Kirstenbosch friend. In 1998 Conservation International identified 17 'mega-diverse' countries in which two-thirds of global biodiversity was

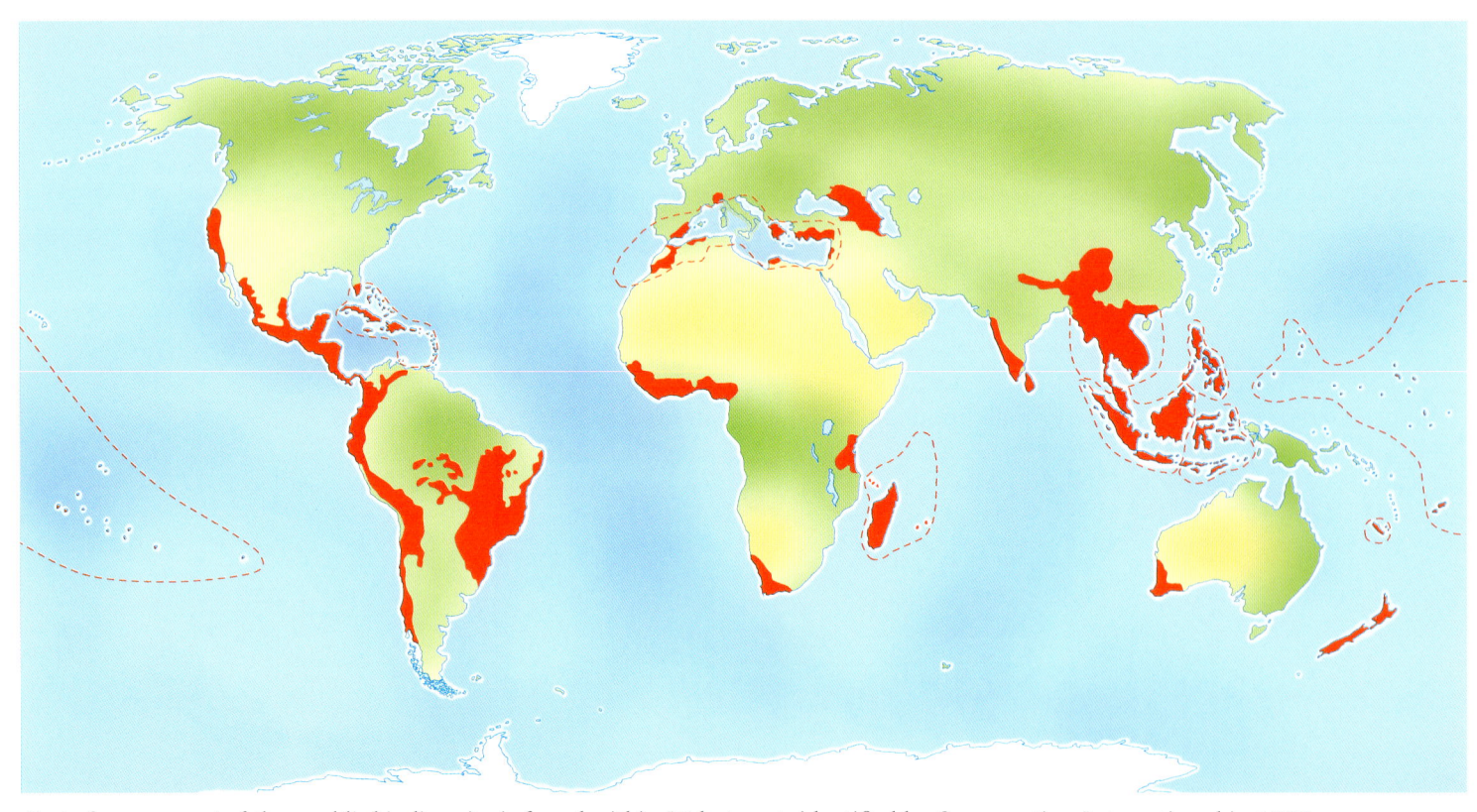

Forty-four per cent of the world's biodiversity is found within 25 hot spots identified by Conservation International in 1998.

70   KIRSTENBOSCH

found – and later identified 25 hot spots in which 44 per cent of the world's vascular plants occur in only 2 per cent of the world's land area. Today, 34 hotspots are recognised, with no fewer than three of these in South Africa – the Cape Floristic Region, the Succulent Karoo, and the Maputaland-Pondoland-Albany Hotspot.

In every debate about the validity and utility of the hot-spot concept, there has been consensus on one issue – that the Cape Floristic Region is the hottest of them all. So what are the hot-spot criteria?

First, a hot spot is assessed in terms of three measures of its species diversity: the number of species within a homogeneous community; the amount of change, or turnover, in species composition across a gradient; and the change in species composition within similar habitats but in different geographical locations, such as on neighbouring mountain peaks. A walk up the mountain backdrop to Kirstenbosch will illustrate how rapidly plant species composition changes, from one small community dominated by proteas, to another dominated by ericas, or to the restio communities in the moist bogs of the summit. If one spends a day listing species on the plateau of the Table Mountain summit, followed by a day in similar mountain fynbos just across the Cape Flats, in the Hottentots Holland Mountains, one might find less than a 50 per cent similarity in the species of these two sites, each of which, at first glance, presents the same rather dull, olive-green heathland appearance.

High biodiversity occurs where all three measures are simultaneously high. The clearly delimited 470 square kilometres of the Peninsula had, at last count, 2 285 species of fern and flowering plant, of which more than 150 occur nowhere else in the world. A comparison can be made with the flora of the United Kingdom, which has an area of

Red List of South African plants

The *Red List of South African Plants 2009* includes the contributions of 169 botanists, brought together by an editorial team competently led by SANBI's Domitilla Raimondo.

The new list sounded a warning: of the 20 456 species assessed, of which 13 265 (65 per cent) are endemic to South Africa, 2 577 species were considered in danger of regional or global extinction. A further 2 232 are listed in other categories of conservation concern, giving a total of 4 809 species, or 23.5 per cent of the flora – one in every four South African plants – in some level of danger.

The good news – if it can be called such – is that of the 59 species believed to have gone extinct in an earlier (1996) survey, no fewer than 18 have been rediscovered in the course of the latest survey programme. But in many cases these rediscovered species are represented by very small populations – representing, in effect, 'the living dead'. The really bad news is that the total number of species listed as Extinct or Possibly Extinct has increased from 59 to 116, a 96 per cent increase since 1996.

The subcontinent's status as a floristic hot spot brings with it the responsibility of safeguarding our heritage, nowhere more so than in the Fynbos Biome: while South Africa as a whole has 2 577 plant species (13 per cent of its flora) considered to be in danger of extinction, no fewer than 1 736 of these are from the Fynbos Biome, with a total of 3 087 (36 per cent) of the fynbos flora considered to be of conservation concern. High levels of land transformation in the Cape lowlands, and the invasion of alien species on both lowlands and in the mountains, are the main causes of species loss.

Thus the high levels of species richness, endemism and the threat of extinction make the tiny Cape Floral Kingdom the 'hottest hot spot' of them all.

The Red List of South African Plants, 2009, *sounds a warning about species loss in the region.*

308 000 square kilometres, with just 1 472 native species, only 20 of which are endemic. Table Mountain National Park, covering an area similar to that of Greater London, has 1 470 species – equal to that of the whole of the UK, in the area of just one city. Even the Amazon Basin, famously rich in species across its many-layered forests, has only one third as many species per 10 000 square kilometres as does the fynbos.

Endemism is the second criterion in hot-spot analysis. Endemics are species (or genera or families) that are found nowhere else on Earth. We have already seen how rich the Cape Floral Kingdom is in endemics, with no fewer than 68 per cent of its 9 381 species being endemic to its 90 760 square kilometres. Local, or narrow, endemics can be found on most mountains or unusual soil types – adding, at fine scale, to the complexity and vulnerability of the flora.

The measures of species richness and endemism of plants within the Cape Floral Kingdom are exceptional, by any standard. But richness and endemism are only two of the criteria used to assess and identify hot spots – threat is as important as the first two.

Since the mid-1970s, South African scientists have collaborated in ongoing projects aimed at measuring and monitoring the levels and kinds of activities and processes threatening the survival of our plant and animal species. Initiated by the then National Programme for Environmental Sciences, co-ordinated by the Council for Scientific and Industrial Research (CSIR), the programme produced an impressive series of *Red Data Books*, which presented syntheses of knowledge on the status of rare plants, birds, mammals, fish, frogs, reptiles and butterflies. In 1996 Kirstenbosch Research Centre conservation scientist Craig Hilton-Taylor, working with a dozen colleagues around the region, listed 4 149 species assessed in South Africa, Botswana, Lesotho, Swaziland and Namibia. Of these, 1 435 were considered globally threatened with extinction at the time.

Hilton-Taylor's database was later vastly expanded by a new Threatened Species Programme, again led by SANBI, but now from its Pretoria offices. From 2002 to 2008, the team compiled the world's most comprehensive synthesis on a national flora ever undertaken – the full 20 456 species of fern and flowering plant of South Africa being assessed using the objectively rigorous 'IUCN 2001' criteria across seven categories of threat.

A gallery of moraeas, all endemic to the Cape Floral Kingdom:
CLOCKWISE FROM TOP LEFT
Moraea aristata CR; M. tulbaghensis EN; M. gigandra EN; two colour forms (mauve and yellow) of M. vilosa LC; M. loubseri CR; and M. calcicola EN; (IUCN categories applicable here: CR – Critically Endangered; EN – Endangered; LC – Least Concern)

The Cape Floral Kingdom 73

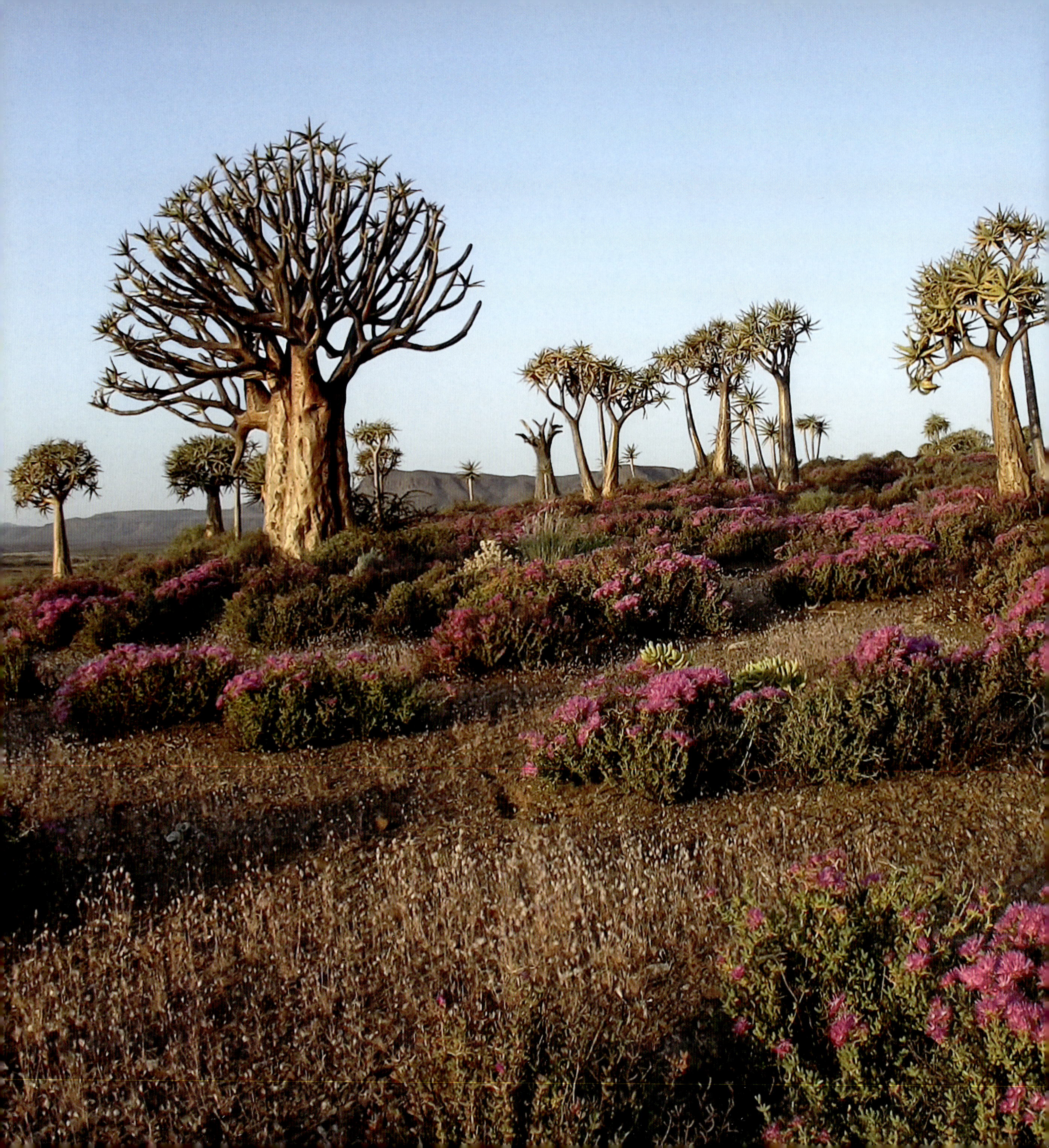

CHAPTER FOUR

The plant hunters

Discovering and documenting the diversity of life

Every district always has something peculiar to itself ... The common saying Semper aliquid novi ex Africa, *should still hold good for many years to come. The pleasure enjoyed by a botanist, who finds all at once so rich a collection of unknown, rare and beautiful vernal flowers, in so unfrequented a part of the world, is easier to be conceived than described ...*

ANDERS SPARRMAN
A voyage to the Cape of Good Hope ... and round the world ... 1772 to 1776

On 2 November 1774, Carl Thunberg and Francis Masson climbed the Bokkeveld escarpment, discovering the amazing Aloe dichotoma, *or Kokerboom, at a site not far from the Hantam National Botanical Garden.*

Early founders of South African botany

Today we know that the Cape has been occupied by humans for some 200 000 years: first by the San, who were hunter-gatherers, and later by Khoikhoi pastoralists. Little is recorded of their original observations on, and uses of, plants. One of the very few depictions of plants in many tens of thousands of rock paintings and petroglyphs is that of a group of hunters with aloes, which are believed to be *Aloe ferox* and *A. broomii*, from a site in the eastern Free State. Little else of botanical detail has survived the steady decay of what were once rich galleries of rock paintings. Fortunately, much of the knowledge of South Africa's first peoples was passed down from the Khoikhoi to Bantu-speaking pastoralists who, in turn, came into contact with, and were employed by, Dutch-speaking farmers occupying ever wider areas of the country in the 18th and 19th centuries. William Burchell, whose travels from 1810 to 1815 covered a vast area of the Cape interior, documented some of the extensive knowledge he gained encountering rural peoples. Many current plant and place names – *kukumakranka*, *buchu*, *karoo*, *ganna* – are derived from the original San and Khoikhoi languages, sadly now long extinct.

This illustration of one of the very few surviving San cave paintings depicting plants shows a group of hunters with aloes, believed to be Aloe ferox *and* A. broomii.

In this chapter we must, inevitably, deal with many 'firsts'. The first person to collect and record details of plants while in the field at the Cape was the missionary Justus Heurnius (1587–1653), who visited in 1624 while on his way from Holland to Batavia. The first professional botanist in the Cape was the German Paul Hermann (1646–1695), who stopped off in 1672 on his way to Ceylon (Sri Lanka) as a ship's medical officer. Hermann built up a herbarium of both Cape and Ceylon plants, and returned to Holland in 1680 to take up the post of Professor of Botany and Director of the famous Botanic Garden at Leiden. His collections, reaching the attention of Swedish botanist Carolus Linnaeus many years later, left the great Swede overwhelmed (see passage quoted at the head of chapter 3).

The botanical and zoological exploration of South Africa really began with Simon van der Stel (1639–1712), Governor of the Cape from 1679 to 1699, who led an expedition to the fabled Copper Mountains of Namaqualand in 1685. This major expedition lasted five months, with many adventures en route. Of interest are the frequent sightings of, and encounters with, large game such as black rhinoceros, elephant, lion, leopard and hippopotamus – pointing to high populations of these species in the Western Cape at the time. The expedition included a German artist, Heinrich Claudius (1655–1697), who documented the animals and plants encountered, and was the most prolific illustrator of the Cape flora during the 17th century.

In 1751 Rijk Tulbagh (1699–1771) became Governor of the Cape. Like Van der Stel, he was an intelligent, progressive personality, interested in natural history, and actively encouraged the exploration of the interior. Despite having arrived at the Cape as a 16-year-old cadet, he must have educated himself in many disciplines, for he was able to conduct an active correspondence with Linnaeus on varied topics – in Latin. Botanical collections prospered during his term of office, but he was to miss the great awakening of Cape plant exploration by just one year, dying as he did in 1771.

The medicinal properties of Aloe ferox *have been appreciated from the earliest times. Watercolour by Eric Judd, from the Kirstenbosch Biennale, 2004*

Earliest documented records of Cape plants

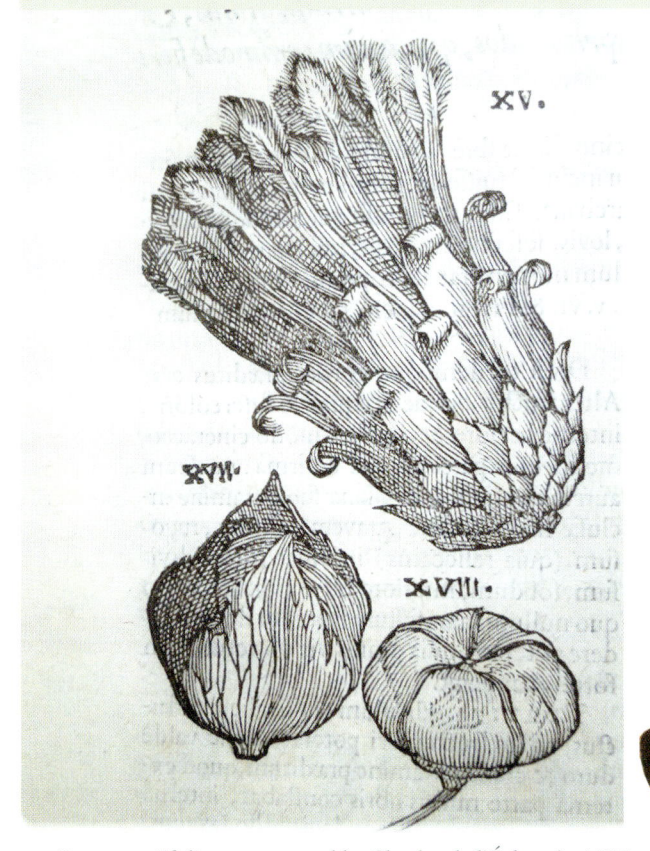

Protea neriifolia *as portrayed by Charles de l'Écluse in 1605*

Protea neriifolia, *painted by Peta Stockton in 2005, four centuries after it was first illustrated by Charles de l'Écluse*

The first documented records of South African plants come from the year 1605. Some years earlier, specimens of two strikingly attractive plants, *Protea neriifolia* and *Haemanthus coccineus*, had been collected in the Cape, and ended up in the hands of two leading botanists of the Low Countries. The Flemish botanist Charles de l'Écluse (1526-1609) – who is credited with the introduction to the Netherlands of tulips from Turkey and the Levant, laying the foundation for the 'tulipomania' bulb cult of the 17th century – illustrated *P. neriifolia*, which he described as 'an elegant thistle', from material gathered on the Cape coast, possibly at Kogelbaai, in 1597. A fellow Flemish student at Montpellier, Mathias de l'Obel (1538-1616), published an illustration of *Haemanthus coccineus* from material probably collected on the slopes of Table Mountain in 1603. Both illustrations, published in 1605, were heavily lined woodcuts (characteristic of the medieval herbals), a genre that was about to be superseded by the beauty and delicacy of copperplate engraving.

The plant hunters

The botanical trio of 1772

The year 1772 should be celebrated as the founding year of South African botany. For this was the year in which Anders Sparrman, Carl Thunberg and Francis Masson arrived in the Cape – independently, but no less fortuitously. Their collections added over 1 500 species to the list of plants known from the Cape.

On 13 April 1772, the Swede Anders Sparrman, a student of the great Linnaeus (see page 85), arrived in Cape Town. Just three days later, on 16 April, another Swede, Carl Peter Thunberg, arrived. On 30 October, the Scotsman Francis Masson arrived. It is improbable that any one land in the new worlds being explored during the 18th century had such a surge of botanical collecting as occurred in the Cape through the last quarter of the 1700s.

Anders Sparrman (1740–1820), like most of the natural scientists of the period, studied medicine and travelled to distant lands as a ship's surgeon (botany was seldom the main thrust of these pioneers' duties, but was taken up as an extra activity). He spent six months at the Cape before joining James Cook on the *Resolution* and passing the next 28 months at sea, circumnavigating the globe, and returning to the Cape for a further 14 months before heading home to Sweden. His detailed observations on the social life, agriculture, natural history and landscapes of South Africa remain one of the classics of the time.

Anders Sparrman (1740–1820)

As much as one admires the fortitude of Sparrman, that of Thunberg is even more impressive. Carl Peter Thunberg (1743–1828) had also studied under Linnaeus and, as a young graduate in medicine, took the opportunity to travel to Japan. However, he first had to learn Dutch because, at the time, Holland enjoyed almost exclusive access to Japan, with other European nations being barred entry. So Thunberg travelled to the Cape in April 1772 to study the Dutch language and customs, continuing to Japan in March 1775, where he remained until his return to Sweden, via Java and Ceylon (Sri Lanka), in 1779.

Thunberg introduced a custom that has become a habit among Cape botanists – mountain climbing. He climbed Table Mountain 15 times during his visit, no small accomplishment given the absence of roads or even paths up the rough slopes. His collections from the Cape included 3 100 species, of which over 1 000 were new to science. He is credited with naming more than 2 300 new species from his collections in the Cape, Japan, Java and Ceylon, leading to the comment – 'God created, Linnaeus ordered, Thunberg described'.

He is commemorated by the genus *Thunbergia* of which the beautiful pale blue, cream-throated *T. natalensis* flowers abundantly in Kirstenbosch. As author of the first comprehensive *Flora Capensis* (1807–1820), he is recognised as the 'Father of South African Botany' – and his *Flora Japonica* should similarly place him as 'Father of Japanese Botany'. He succeeded Linnaeus to the Chair of Botany at Uppsala in 1784, which he held until his death in 1828.

Perhaps no greater tribute could be offered to Thunberg than the words of Peter MacOwan, writing in 1886:

> '... as long as in our paradise of flowers there wanders a single botanist, so long will the name of Thunberg be held in honoured remembrance.'

Carl Peter Thunberg (1743–1828)

KIRSTENBOSCH

The third of the trio of 1772 was Francis Masson (1741–1805). Significantly, he was born in Aberdeen – it is said that the custom in England at the time was to select thrifty Scots as gardeners. Such was the case at Kew, and Sir Joseph Banks recommended to King George III that Masson should be sent to the Cape to collect interesting plants for the Royal garden. He arrived in the Cape on Cook's *Resolution* on 30 October 1772, just as Sparrman was setting off for the South Seas. Unlike Sparrman and Thunberg, Masson was a gardener, not a scientist. This, in effect, added to his value. His horticultural training resulted in his selecting plants of gardening merit, rather than mere scientific curiosity. He introduced 400 new species into horticulture, including 50 *Pelargoniums*, the source material of the profusion of 'geranium' hybrids and cultivars that have added so much colour to homes and gardens around the world. Other genera that he introduced included *Ericas*, *Cinerarias*, *Ixias* (including the remarkable green *Ixia* – *Ixia viridiflora*) and, to the delight of King George III, *Strelitzia reginae*, named after Charlotte of Mecklenburg-Strelitz, queen-consort.

Francis Masson (1741–1805)

Masson left the Cape in March 1775 – just as Sparrman arrived back – and returned to Kew, where his introductions caused a sensation. Soon after his return he resumed his travels, collecting specimens in the Azores, Madeira, the Mediterranean, the West Indies and, in 1786, in the Cape again, for another nine years. Such was the indefatigable nature of the plant collectors of the 18th century!

Erica massonii, *a striking heath, was named after Francis Masson, the great collector of the Cape's horticultural gems.*

The plant hunters

This vista of Kirstenbosch offers a visual feast from foreground to distant horizon, thanks to the extraordinary and unparalleled beauty of the Garden's setting.

80 KIRSTENBOSCH

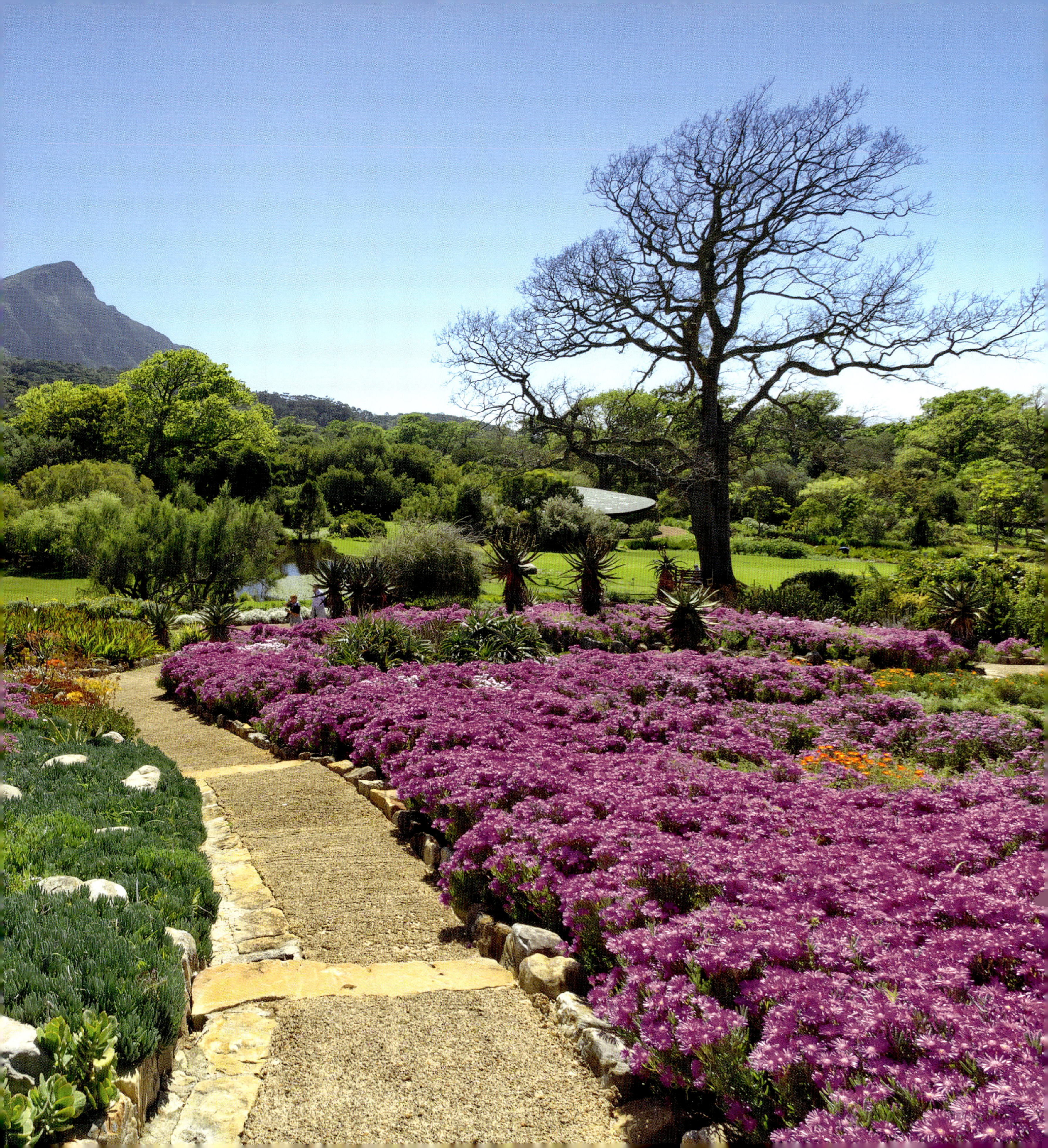

A passion for Cape plants

The closing years of the 18th century and the first decades of the 19th saw many more important collectors in the Cape, as the fame of its floral wealth spread through the gardens of Europe. Among them were the Scot William Paterson, collector for the Countess of Strathmore; Germans Frans Boos and Georg Scholl, collectors for Emperor Josef II; another Scot, James Niven, collector for London businessman George Hibbert and later for the French Empress Josephine; the Austrian Ferdinand Bauer – one of two brothers, both botanical artists extraordinaire; the German Martin Lichtenstein, who also served as family doctor to the Governor at the Cape; the English scientist and explorer William Burchell; another collector from Kew, James Bowie; and a trio of German collectors – Carl Zeyher, Christian Ecklon and Franz Drege – who together despatched more than 200 000 specimens of 8 000 species to the herbaria of Europe. These were men of the Enlightenment, who pursued truth with passion and unfettered enthusiasm. Each made impressive contributions to our knowledge.

The surge of collecting activity at the Cape was not always a free-for-all scramble. As early as 1805, collecting permits were needed. In the winter of 1804, Empress Josephine Bonaparte, keenly interested in obtaining plants from the Cape of Good Hope for her garden at Malmaison, despatched James Niven to the Cape for his second visit. Niven arrived in Table Bay in early 1805 and, on 3 April, was issued a permit to collect plants 'in the service of Her Majesty the Empress of the French' – probably the first collecting permit on record in Africa, although the practice soon fell away as colonial interest waned. The introduction of permits to monitor or restrict collecting had perhaps been triggered by the visit in the 1770s of Scotsman William Paterson, who was suspected of being on a spying mission. Paterson went on to Australia, making major collections there and even becoming, for a brief period, Lieutenant Governor of New South Wales.

Streptocarpus kentaniensis is a recently described species from the genus discovered in 1826 by Kew collector James Bowie. (Watercolour by Vicki Thomas)

William Paterson, collector and colonial administrator, after whom the Endangered Erica patersonii *was named*

Erica patersonii was named for the Scottish collector William Paterson who visited the Cape in the 1770s.

82 KIRSTENBOSCH

Gunn and Codd – South Africa's botanical historians

The individual endeavours, adventures, sacrifices, disasters, triumphs and, in some cases, tragedies, of South Africa's botanical pioneers are told in the scholarly volume *Botanical Exploration of Southern Africa*. This monumental work was compiled by Mary Gunn and Leslie Codd of the then Botanical Research Institute, Pretoria, and published in 1981. Of this volume, John Rourke, himself a scholarly writer on the history of South African botany, commented 'every now and then there appears in South African botanical literature a book of exceptional distinction, a book which even from the moment of its appearance is recognized as a classic'.

Mary Gunn (1899-1989) – Miss Gunn, as she was known to all – was a remarkable woman who commenced her career in 1916 as a clerical assistant to I.B. Pole Evans (see page 44). After retiring in 1954, Miss Gunn continued active work in the library (later to be named after her), which she built up to become the most comprehensive botanical library in Africa, if not the southern hemisphere; it is one of SANBI's most valuable, if little appreciated, assets. Always ready to assist young students, and a virtual 'walking encyclopaedia' on African botanical literature, Miss Gunn inspired several generations of botanists to look beyond current publications.

Leslie Edward Worstal Codd (1908-1999), Director of the Botanical Research Institute from 1963 to 1973, left a lasting legacy to South African botany and was another source of great encouragement to young students.

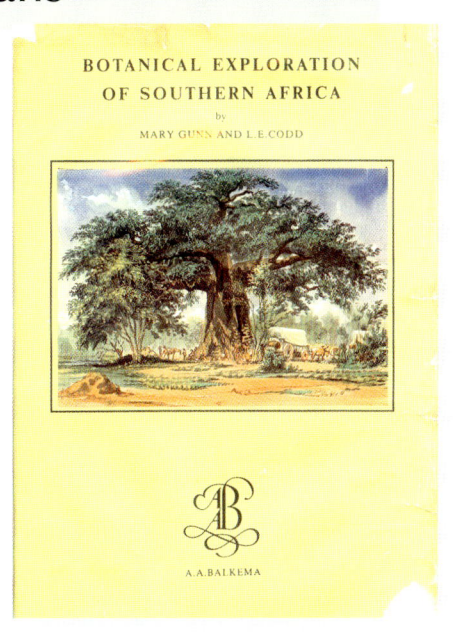

Mary Gunn and Leslie Codd's Botanical Exploration of Southern Africa *was a classic of its time.*

James Niven, represented here by Nivenia stokoei, *had to obtain a collecting permit in 1805 from the Batavian Government in the Cape, perhaps following suspicions that an earlier collector, William Paterson, had engaged in spying activities at the Cape in the 1770s.*

The plant hunters 83

South African collectors at the Cape

This account of the collectors would not be complete without reference to several major contributors to our knowledge of the fynbos flora during the last century. During the late 19th century, botanists born, or at least resident, in South Africa started replacing visiting collectors as the discoverers of new species. A founding figure for many of the institutional developments in Cape botany was an Englishman who came to South Africa as a young man to set up business in Grahamstown. Harry Bolus (1834–1911) was a stockbroker-turned-botanist, a mountaineer and a philanthropist. He not only collected widely throughout the country, but also published a three-volume work on the orchids of South Africa – *Orchidearrum Austro-Africanarum Extra-Tropicum* – and personally drew and coloured many of its plates. In 1902 he founded a Chair in Botany at the South African College, later to become the Harry Bolus Chair of Botany at the University of Cape Town (UCT); and he bequeathed his herbarium, library and most of his fortune to the university on his death in 1911. His herbarium was housed at Kirstenbosch from 1924 until its transfer to UCT in 1953. It remains one of the most important herbaria in the country.

Harry Bolus (1834–1911), stockbroker-turned-botanist, mountaineer and philanthropist

We have already spoken about Rudolf Marloth, the pharmacist/analytical chemist who founded our knowledge of fynbos ecology and collected over 15 000 specimens of Cape plants. Another mountaineer of unceasing energy, who added many species from the highest peaks of the Cape mountains, was the Yorkshire-born Thomas Pearson Stokoe (1868–1959). Stokoe collected over 16 000 specimens – and celebrated his 91st birthday collecting plants in the Hottentots Holland Mountains with the then Kirstenbosch Director Brian Rycroft. Stokoe is remembered in the 30 species now carrying his name.

Yorkshire-born Thomas Pearson Stokoe (1868–1959), mountaineer and collector of plants in the Kogelberg

Yet a further passionate mountaineer, and prolific plant hunter/collector – she amassed some 37 000 specimens – was Elsie Esterhuysen (1912–2006), who discovered over 150 taxa, and after whom a staggering 56 species have been named. Elsie was a familiar sight in Kirstenbosch until her nineties, when she could be seen racing up Skeleton Gorge as if still a teenager.

Such, then, is a sampling of some of the many pioneers of South African botany. To them we owe South Africa's status of possessing one of the most comprehensively collected, researched and documented floras on Earth.

By 1980 the rate of new specimens accessioned at the National Herbarium, Pretoria, reached a peak of 18 000 per year, reflecting the local enthusiasm for botanical collecting. Of concern is that, by 2010, the number of new accessions had dropped to 2 000 per year, a level last recorded in 1905. The transfer of energies from taxonomy to ecology has much to do with this, but an underlying cause of the demise of field collecting is the increasing weight of bureaucracy and legislation attending the simple practice of biodiversity inventory. The United Nations Convention on Biodiversity (CBD), for all its impressive outcomes, is victim of the unintended consequences of its policies on the control of access to biological material. Legislation has now so constrained collecting activities that the CBD might soon be recognised as the 'dead hand on biodiscovery'.

Elsie Esterhuysen (1912–2006), discoverer of over 150 taxa and after whom a staggering 56 species have been named

Taxonomy – the science of classifying and naming organisms

Until 1753, plants and animals were collected, preserved and described using long and often confusing, or difficult-to-catalogue, names – and all this in Latin, the universal language of the day. By the early 18th century, plant and animal collections resulting from the exploration of Africa, Asia and the Americas were burgeoning, and there was an urgent need for an efficient and consistent system of classifying and naming the thousands of new specimens arriving in the museums, herbaria and universities of Europe. The science of classifying and naming organisms, known as taxonomy, was born. Taxonomy includes four components – classification; naming (nomenclature); the circumscription and description of species, genera, families, etc; and the production of identification tools. Closely related is the science of systematics, the study of the diversity of organisms and the relationships among them, which includes elements of evolution, phylogeny, population genetics and biogeography. Today the lines between taxonomy and systematics have merged as the implementation of molecular technologies has called for a greater synthesis of approaches.

Carolus Linnaeus

The father of modern taxonomy was the Swedish naturalist Carolus Linnaeus (1707-1778). Linnaeus revolutionised plant taxonomy with his so-called 'sexual system', based on the number and arrangement of the reproductive organs, the stamens and pistils. Linnaeus created a large number of genera and assigned each species to a genus. Today, taxonomists group genera into families, families into orders, orders into classes and classes into divisions within the Plant Kingdom.

It is certainly thanks to the great Linnaeus that we have an orderly arrangement of our knowledge of global biodiversity. Few words can better express an appreciation for the Cape's flora, than those in a letter from Linnaeus to Governor Tulbagh written in the mid-18th century:

The father of modern taxonomy: Swedish naturalist Carolus Linnaeus (1707–1778)

'May you be fully aware of your fortunate lot in being permitted by the Supreme Disposer of events to inhabit, but also enjoy the sovereign control of that Paradise on Earth, the Cape of Good Hope, which the Beneficent Creator has enriched with His choicest wonders. Certainly if I were at liberty to change my fortune for that of Alexander the Great, or of Solomon, Croesus or Tulbagh, I should without hesitation prefer the latter.'

The plant hunters

Cape stock-rose Sparrmannia africana

Dwarf Thunbergia Thunbergia natalensis

Wild pomegranate Burchellia bubalina

A 'species' is the most fundamental level of classification – it is the building block of any analysis of biodiversity. In its simplest definition, members of a species can interbreed with one another, but not with other species. Next up the hierarchy of taxonomy is the genus – a grouping of similar, closely related species. The genus is a somewhat subjective category, leading to much debate among taxonomists and to confusion for those foreign to the arcane principles of taxonomy, or to the bizarre International Code of Botanical Nomenclature. Essentially, a genus is the ensemble of its component species and, likewise, a plant family is the ensemble of its component genera.

The name of a species consists of two parts: the name of the genus, followed by the term peculiar to the species, called the specific name or epithet. The genus name is comparable to a surname, and the specific name to a first name, although they are subject to a few rules: the genus, with an initial capital, comes first, the species (all lower case) comes second, and the two names (forming a binomial) are printed in italics (or a style that distinguishes them from the surrounding text). We can use the scientific name of South Africa's national flower, the King Protea, *Protea cynaroides*, to illustrate these points.

Taxonomists often use features of plants when naming them. The genus name *Protea* was reputedly chosen by Linnaeus after the Greek god Proteus, who was able to take many forms – in reference to the very variable genera of the family Proteaceae. The specific epithet *cynaroides* means 'looking like a Cynara' – the Globe Artichoke. Linnaeus also introduced the custom of naming genera and species after places or people, most frequently a fellow botanist, collector or patron. Thus we have names like *Sparrmannia africana*, *Thunbergia natalensis* and *Burchellia bubalina*.

A brave attempt to compile a new *Flora of Southern Africa*, initiated in 1963 by Robert Allan Dyer, Director of the BRI from 1944 to 1963, and the team of taxonomists at the National Herbarium, has lacked the support needed to mobilise and complete such an ambitious project. But this has not prevented the compilation of an inventory of our flora. From Linnaeus' 1759 list of 502 species, Thunberg's 1820 list of 3 100 species, and Harvey's 1838 list of 7 860 species, in 1874, Pappe extrapolated the flora of South Africa as a whole to be 18 000 species. The most recent checklist compiled in 2009 comprises 20 564 species, excluding introduced and naturalised aliens.

The Cape flora remains the most exhaustively documented in Africa, with the Compton Herbarium's John Manning and his Missouri Botanical Gardens colleague, Peter Goldblatt, producing two editions of the landmark catalogue of *Cape Plants* (first published by Pauline Bond and Peter Goldblatt in 1984), a volume that has become an indispensible tool for taxonomists, ecologists and conservationists.

An outstanding botanical feat

It was the Irishman William Henry Harvey (1811–1866), residing in Cape Town during three periods between 1835 and 1842, who set the benchmark for detailed work on the region's flora. When appointed Colonial Treasurer to the Cape at 24 years of age, Harvey drew on the skills he had learnt in his father's merchant business. He was a passionate botanist without formal training, using his leisure time to prepare *The Genera of South African Plants*. It was printed in Cape Town in 1838 – the first substantial botanical book published in South Africa. This work, accomplished in less than four years by a young, self-taught botanist, fully occupied with the duties of Colonial Treasurer, is in itself quite outstanding – a virtually unsurpassed feat of botanical productivity. Harvey catalogued 7 860 vascular species in his *Genera*. Remarkably, his original copy, with many hand-written notes, is now in SANBI's Mary Gunn Library in Pretoria. He wrote the book in the hope that it would 'be sent to resident doctors, clergymen, etc. scattered about the country to excite their idle minds to send specimens to Cape Town'. Having returned to Europe in 1842, he was appointed to the Chair of Botany at the University of Dublin in 1856. Together with Otto Wilhelm Sonder (1812–1881) of Hamburg, Harvey then proceeded on his *Flora Capensis* project, to which he contributed the major part of the first three volumes in 1860, 1862 and 1865 – before ill health and his death in 1866 curtailed the project. Fortunately for South African botany, the project was continued, after several decades, under Sir William Turner Thiselton-Dyer at Kew, and completed in 1933. This was the same Thiselton-Dyer who had earlier encouraged Pearson to go to the Cape (see page 37).

William Henry Harvey (1811–1866), Colonial Treasurer and self-taught botanist

Not only has Manning contributed to the scientific documentation of the Cape's flora, he has produced over a dozen richly illustrated handbooks and field guides to the country's flora, making it accessible to all who share his passion for the region's wild flowers.

As our knowledge of the South African flora has expanded, so too has that of the African flora as a whole. In 1992, Kirstenbosch hosted a meeting of African botanists to plan a co-operative project aimed at building regional capacity in plant taxonomy and herbarium management. The project, which became known as SABONET (Southern African Botanical Diversity Network), succeeded in training over 30 botanists at postgraduate level, rehabilitated over a dozen herbaria, and computerised the information on some 200 000 herbarium specimens. Building on SABONET was the African Plants Initiative (API), driven by the enormous energies of Gideon Smith, Head of Systematics Research in SANBI, and by teams of taxonomists across Africa and in Europe and the USA. With generous support from the Andrew Mellon Foundation, the API published, in 2006, the first comprehensive checklist of the 50 000 plants known in sub-Saharan Africa – one of the many impressive outputs of the SABONET and API projects for which SANBI can be justifiably proud.

Goldblatt and Manning's Cape Plants *– a modern benchmark in floristic inventory*

The plant hunters

Estimates of a total world flora of 250 000 species have increased to 370 000 and to a possible 400 000 flowering plants and ferns across the globe. Such estimates change the baseline of the popular statistic given in conservation publications that South Africa is home to 10 per cent of the world's plant species. More accurate estimates reduce this to 5 per cent, but given that 65 per cent of these occur nowhere else on the planet, the importance of ensuring their survival for the benefit of humanity remains a high priority.

The Linnaean system predated Darwin's theory of natural selection, and viewed species as being static entities – natural beings that had been created by God, and were unchanging. This fundamental flaw in his system has not prevented it from surviving as the most basic element of all research and knowledge management in the biological sciences for over two and a half centuries. Despite its shortcomings, it is here to stay!

ABOVE *SABONET team members show off a wall of plant presses from a collecting expedition on the Nyika Plateau, Malawi.*

FAR LEFT *Type Specimen of the Marsh Rose* Orothamnus zeyheri *(see also photograph on page 226) was collected by Carl Zeyher in 1846 and named by Carl Pappe in 1848. It was first deposited in the South African Museum and transferred to the Compton Herbarium in 1956.*

LEFT *Holotype of* Erica penduliflora *was collected in 1999 by Ted Oliver of the Compton Herbarium and named by him in 2001.*

88 KIRSTENBOSCH

The artists – illustrators for science and society

South African botany has been served by a succession of highly skilled artists and illustrators, too many for more than a cursory mention of just a few of them here. The Austrian Frans Bauer (1758–1840), one of the famous botanical artist brothers and perhaps the greatest illustrator of South African plants, never visited the country. As botanical artist to King George III, he had access to freshly grown material from the glasshouses of Kew, cultivated from seeds and bulbs sent back to London by Francis Masson. Described by the great botanical art historian William Stearn as having come 'nearer to perfection in this field of art and scholarship than anyone before or since', Bauer illustrated hundreds of South African plants 'to their greater perfection'.

John Frederick William Herschel (1792–1871) and Margaret Brodie Herschel (1810–1884) were an extraordinarily talented couple. During their four-year sojourn at the Cape, the famous astronomer drew fine line drawings of some 300 species, each painted with great sensitivity in watercolours by his wife. These paintings, after being inaccessible for a century and a half, have been made available to all through the sumptuously published *Flora Herscheliana* authored by John Rourke, for many years Curator of the Compton Herbarium at Kirstenbosch, and Brian Warner, Professor of Astronomy at the University of Cape Town.

Botanical art flourished in South Africa throughout the 20th century, with work of outstanding quality contributed by Auriol Batten (1918–), Ellaphie Ward-Hilhorst (1920–1994), Thalia Lincoln (1924–), Fay Anderson (1931–), Vicki Thomas (1951–), Gill Condy (1952–) and many more. The fine works of these artists enrich the many monographs on Cape plants produced by botanists at Kirstenbosch, notable among these being John Rourke's *Mimetes* illustrated by Thalia Lincoln, John Manning's *Gladiolus* illustrated by Fay Anderson and Auriol Batten, and Ernst van Jaarsveld's *Plectranthus* illustrated by Vicki Thomas. Botanical artists, including Mary Maytham Kidd and Auriol Batten, spearheaded the surge of field guides in the 1950s and 1960s, before photography replaced them towards the end of the century.

It was not long before works from many of South Africa's rapidly expanding corps of artists found their way into the Shirley Sherwood collection, to be exhibited in leading galleries around the world and in the Shirley Sherwood Gallery in the Royal Botanic Gardens, Kew.

Frans Bauer was botanical artist to King George III, illustrating the plant treasures arriving at Kew from the far reaches of the Empire, such as Erica massonii, *named for the 18th century collector of Cape plants. (Courtesy of the Royal Botanic Gardens, Kew)*

The plant hunters

The Kirstenbosch Biennales

Botanical artists have, for most of the past century, been notoriously poorly paid. What has brought significant change to the fortunes of botanical artists, however, was a change in the market. Much of the credit for this must be given to Dr Shirley Sherwood, a botanist herself, with a keen eye for talent. Dr Sherwood first visited Kirstenbosch in 1990, and was so impressed by the beauty of its flowers, that she immediately sought out artists who might have works for her to purchase for her private collection.

Kirstenbosch was host to a massive exhibition of 130 of the Shirley Sherwood Collection's choicest works in 1996. Inspired by this exhibition, Merle Huntley initiated and curated the first four of the now ongoing *Kirstenbosch Biennales*, which present the best of South African botanical art to an expanding audience. Prices for 'flower paintings' sold at the Biennale now reach previously unheard-of prices. Paintings are judged not only on their technical and scientific merit, but also on their 'wall appeal' – a criterion that has liberated botanical illustration from dusty archives to the galleries of major institutions and private collections. In the case of South African botany, the professional artist has never had it so good.

Covers of Kirstenbosch Biennale *catalogues*

Art works from the Kirstenbosch Biennales: OPPOSITE PAGE, FAR LEFT Strelitzia nicolai *(Ann Schweizer);* LEFT Protea neriifolia *'Kirstenbosch' (Peta Stockton);* THIS PAGE, CLOCKWISE FROM TOP LEFT Gardenia globosa *(Barbara Pike);* Strelitzia juncea *(Miriam Stern);* Kigelia africana *(Ann Schweizer);* Diaphanathe xanthopollinia *(Sibonelo Chilisa);* Protea neriifolia *(Peta Stockton);* Agapanthus walshii *(Vicki Thomas)*

The plant hunters

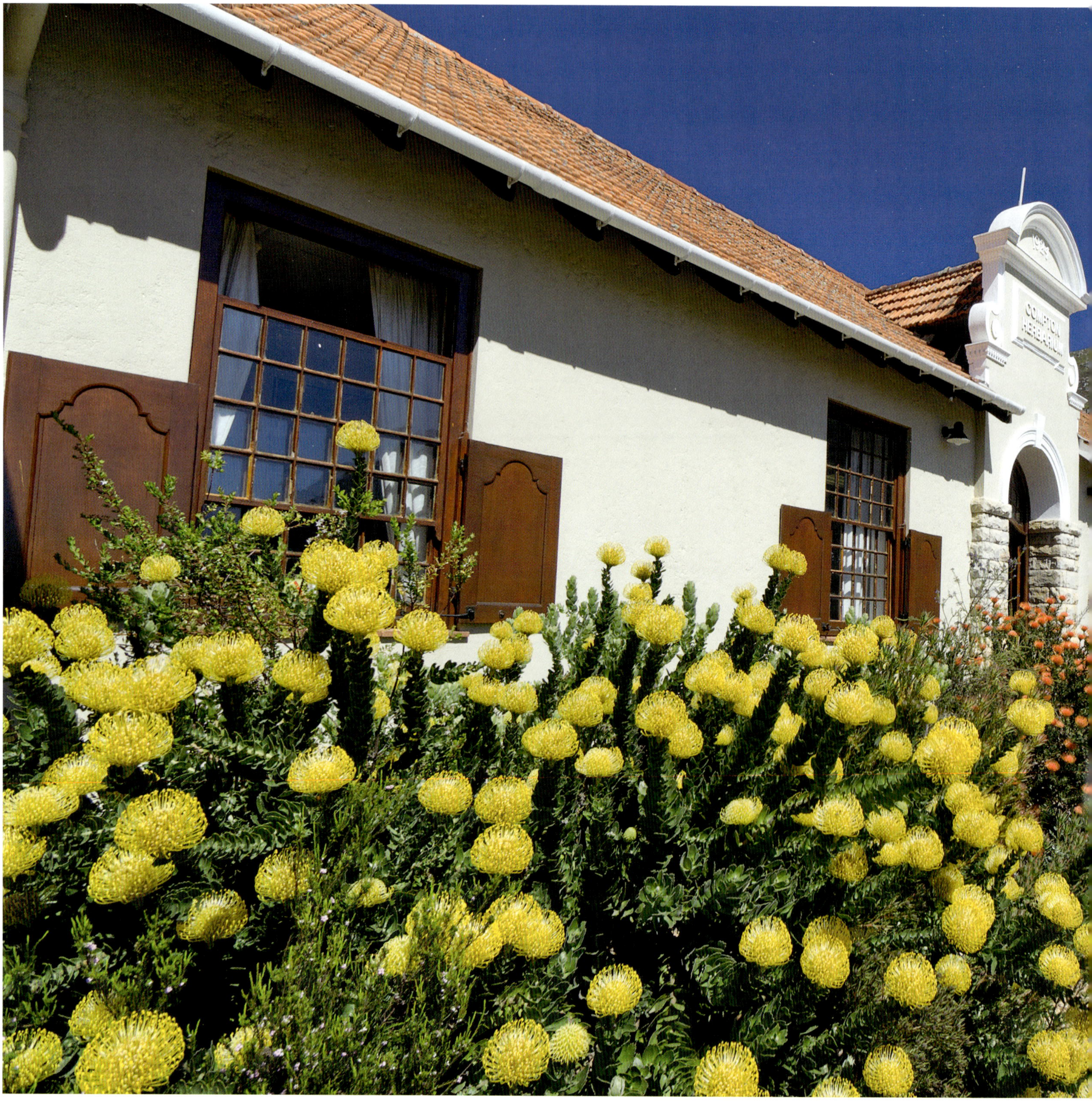

Adam Harrower

The Compton Herbarium

When Harry Bolus died in 1911, he bequeathed his great herbarium, library and most of his personal fortune to the South African College. Had he lived just a few more years, he might well have left his estate to Kirstenbosch – or at least that is what Compton believed. But neither timing nor luck was on Kirstenbosch's side in this matter, for, although the Bolus Herbarium was housed at Kirstenbosch for 11 years (1923–1934), its future lay with what became the University of Cape Town.

The matter remained a source of much concern to Compton, who described the removal of the Herbarium as 'both retrograde and catastrophic'. The reason is simple: a herbarium is the key repository of botanical knowledge that any botanical garden must have if it is to function as a scientific institution. Compton had been trained as a taxonomist and systematist; his interest lay with discovering plants new to science, carefully describing them, classifying them, and curating the dried specimens in a well-organised 'library of plant specimens'– which is what a herbarium is.

Undaunted by the departure of the Bolus Herbarium, Compton set about building a herbarium at Kirstenbosch. His efforts saw the establishment of a Botanical Assistantship in 1933 – until which time he had been the only botanist on the Garden's staff – and by 1940 he had three botanists, a growing library and 9 908 mounted herbarium specimens in 32 cabinets. The trauma of losing the Bolus collection, and the establishment of a competing, well-endowed National Herbarium in Pretoria in 1923, was eased. He later recalled:

'This successful beginning of the new Herbarium gave great encouragement to the belief that the Gardens were after all by nature a definitely botanical centre, and did much to heal the wounds that fate had inflicted upon them'.

Two wings were added to the Compton Herbarium in 1958. Since the transfer of the Herbarium to the Kirstenbosch Research Centre in 1996, the building has served as offices of the Garden Curator.

Within the herbarium's first 10 years it had incorporated over 40 000 specimens in its collection. Much credit for this is due to Winsome 'Buddy' Barker, who, as first Curator of the herbarium from 1939 to 1972, set about establishing the fine standards of its collections. In 1956, a major windfall occurred with the transfer, on permanent loan, of the very valuable collections of the South African Museum to Kirstenbosch. Brian Rycroft, recently appointed Director to the NBG, played a key role in securing the collections for Kirstenbosch, rather than seeing them despatched to Pretoria. The 118 cabinets from the Museum were housed in two new wings that were added to the building in 1958. At the opening of the extensions, the Trustees named the building and its valuable collections the Compton Herbarium – due honour being accorded the champion for systematics at Kirstenbosch over nearly 40 years.

During the 1960s and 1970s, growth of the Compton Herbarium's collections and research activities focused mostly on the family Proteaceae, led by John Rourke, who had been appointed Curator in 1972. But limited funding kept the science programme at Kirstenbosch at a modest level until the 1980s, when Kobus Eloff was appointed Director. Taxonomy was still poorly supported until the 1990s, when a variety of initiatives were energetically pursued. First of these was the building of the new Kirstenbosch Research Centre (KRC), to which the Compton Herbarium was transferred in 1996. The Compton collections were further supplemented in 1996 by the transfer of over 120 000 specimens from the Stellenbosch Herbarium to the new facilities at the Kirstenbosch Research Centre.

Jamesbrittenia bergae – *perhaps the most spectacular ornamental plant recently discovered in South Africa – demonstrates the importance of continued field exploration in the country.*

Pelargonium peltatum, *the South African species from which thousands of hybrid 'geraniums' have been bred, adds colour to the window boxes of Europe.*

There has always been some reservation about expanding beyond the 'comfort zone' of classical alpha taxonomy of Cape plants at the Compton Herbarium. Given that the Cape flora is greater than that of the majority of countries anywhere on Earth, this focus is sensible and has achieved impressive results. Two of the major families – Proteaceae and Ericaceae, have enjoyed special attention, as have the exceptionally rich bulb families – Iridaceae and Amaryllidaceae. Major monographs have appeared on these groups from the Compton team.

Collaboration with other African herbaria had been limited until the 1990s, when the SABONET project provided an opportunity for South African botanists to actively engage with, and mentor, colleagues from a dozen other African countries. The Compton has also made a meaningful contribution to some of the regional and global programmes led by the National Herbarium in Pretoria. The current Curator of the Compton Herbarium, Koos Roux, has been active in cataloguing the fern flora of Africa and its islands. The Herbarium staff have participated in the African Plants Initiative (API) – a massive and successful project, funded by the Andrew Mellon Foundation, which has prepared electronically accessible digital images of the majority of the 'type specimens' of Africa's 50 000 plant species. A type specimen is the reference specimen, usually the first specimen collected, of a new species when it is described and the description published by a taxonomist. Such early reference specimens are of great importance to classical taxonomic research, hence the value of the South African Museum Herbarium, which includes many of the type specimens collected by the early botanists at the Cape.

Researchers examine plant specimens in the Compton Herbarium.

Sharing the knowledge

In 1980, Kay Bergh, one of Kirstenbosch's staunchest supporters for several decades, suggested that the Botanical Society produce and publish inexpensive field guides to the wild flowers of the different floral regions of the Cape. Few of the Botanical Society's many important projects have been more successful. Twelve *South African Wild Flower Guides* had been published by 2011, with over 100 000 copies being sold. Each of these richly illustrated guides has enjoyed active support, and often authorship, by Kirstenbosch botanists and horticulturists. These guides, together with the vast array of other natural history publications, make the Botanical Society's Bookshop at Kirstenbosch one of the most important resources of knowledge available to the general public on the country's biodiversity.

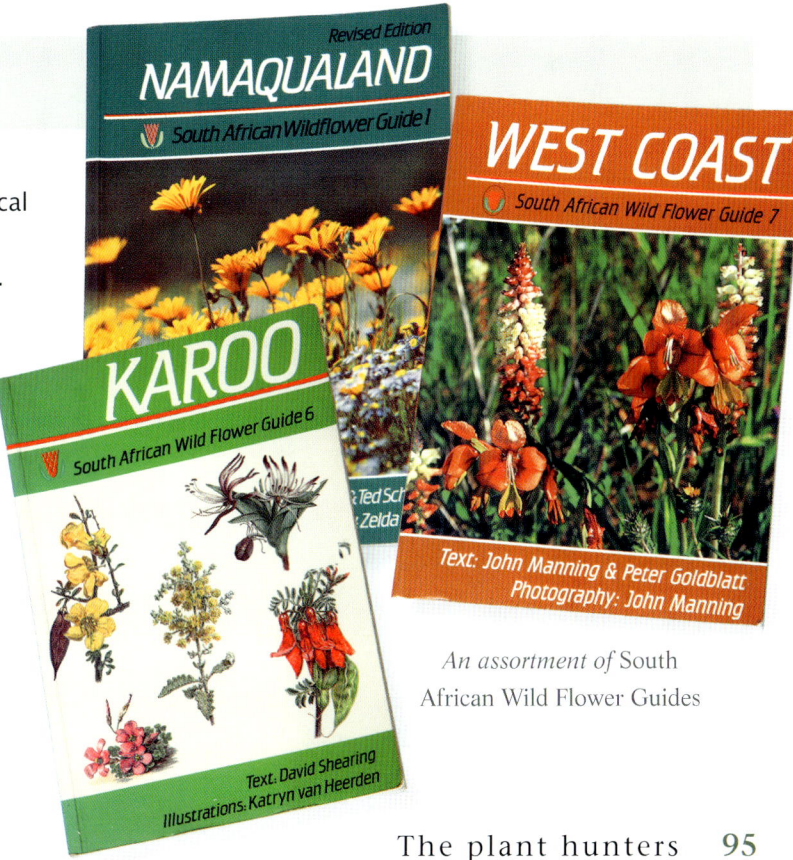

An assortment of South African Wild Flower Guides

The plant hunters 95

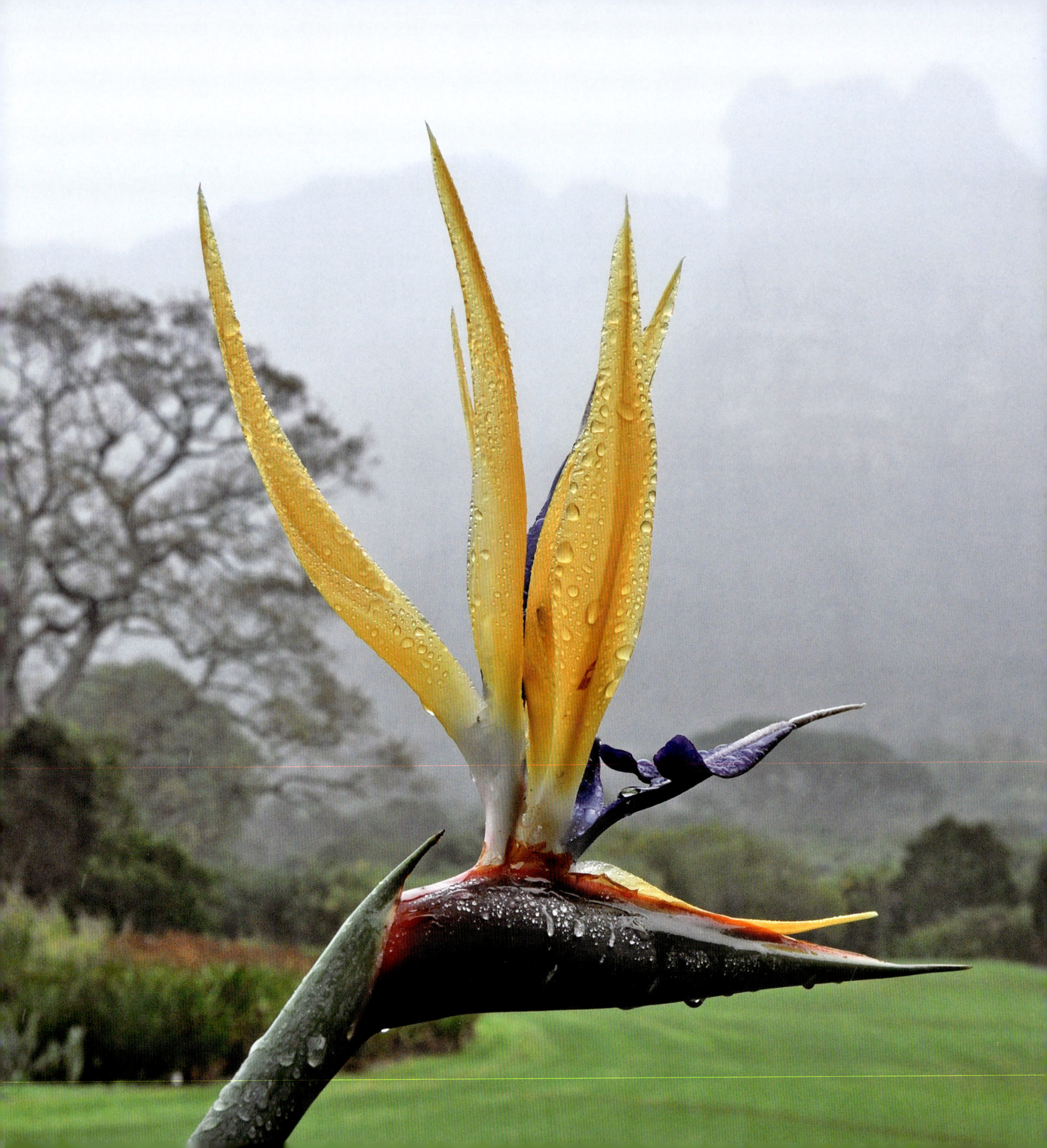

CHAPTER FIVE

New directions in a changing South Africa

The National Botanical Institute years

'The top-management will have to break from the past of the two organisations, and out of the best of both build a new organisation, which will be a national organisation, or at least have the chance to become one ...'

Consultants' Report to the
NBG Board of Trustees, 1989

Kirstenbosch honoured President Nelson Mandela by naming a spectacular colour form of Strelitzia reginae *'Mandela's Gold'.*

The 1990s – a decade of change

On 11 February 1990, Nelson Mandela was freed after 27 years of imprisonment, and a new era began for South Africa. Thirty years after Harold Macmillan's historic 'Winds of Change' speech in the Houses of Parliament in Cape Town, democracy was being born in a country long shunned by the world. The 1990s saw rapid, exciting and massive changes in all aspects of the country, changes that had been in progress over several years as old paradigms began to give way to new.

The establishment of the National Botanical Institute (NBI) through the amalgamation of the National Botanic Gardens (NBG) with the Botanical Research Institute (BRI) was an example of the spirit of change sweeping the land. The new organisation had to break from deeply ingrained dichotomies between the BRI (Pretoria, conservative, well funded, research oriented and perceived to be predominantly Afrikaans speaking) and the NBG (Cape Town, liberal, resource starved, garden oriented and predominantly English speaking). The creation of the NBI, whose Board now reported to the Minister of Environmental Affairs and Tourism, provided for an expanded budget and additional professional and support posts. The first post to be advertised was that for a new Chief Executive Officer to lead the amalgamation and restructuring of the new organisation. The Board of the NBI was guided in its selection by the consultants who had led the BRI/NBG amalgamation process. They laid emphasis on a new kind of leadership – a break from the academic traditions of the NBG and the BRI – and a change to a business-management orientation.

The Board wanted someone free of BRI or NBG loyalties or prejudices, a manager rather than an academic. Their choice was Brian John Huntley. Born in Durban on 20 February 1944, Huntley went on to study at the University of Natal. He spent 15 months on the Prince Edward Islands as botanist to the first Biological-Geological Expedition to the remote sub-Antarctic islands, and worked as ecologist to the Angolan National Parks from 1971 to 1975, before returning to South Africa and to the position of manager for Ecosystem Programmes of the CSIR and, later, the National Research Foundation.

In January 1990 Huntley was appointed CEO of the National Botanical Institute. His links with Kirstenbosch dated from 1960, when he had won the J.W. Mathews Floating Trophy – a national essay competition for schools initiated by Kirstenbosch in honour of its first Curator. The topic was invasive alien plants and their impact on indigenous vegetation. His first research paper, in 1965, appeared in the NBG's *Journal of South African Botany*. He had also worked as a student at the BRI's Natal Herbarium, and for several years had offices in the National Herbarium, Pretoria. Both organisations enjoyed his respect.

Huntley's connection with Kirstenbosch dated from his school days in Durban, when he won the J.W. Mathews Floating Trophy in 1960, 30 years before his appointment as CEO of the NBI.

The most famous visitor to Kirstenbosch during Huntley's tenure was President Mandela, enjoying the Garden that he had first visited as a student.

Mandela's Gold

With the establishment of the National Botanical Institute in 1990, a new logo was sought that would reflect the unique beauty of South Africa's botanical treasures. The bird-of-paradise *Strelitzia reginae* – an endemic of the east coast – was chosen. First collected and dispatched to Kew by Francis Masson in 1774, it had soon become a world-wide garden favourite.

Following the inauguration of President Mandela on 31 May 1994, NBI management proposed that a very special plant be named after the first President of the new democracy. Many recommendations were received, but the choice of a rare yellow colour form of *Strelitzia reginae* won the day. Kirstenbosch Curator John Winter had collected the plant in the 1970s; through careful selection, a cultivar with a splendid golden flower was produced.

Kirstenbosch was honoured when President Mandela paid a special visit to the Garden on 21 August 1996; he was presented with a portrait of 'Mandela's Gold' painted by SANBI artist Gillian Condy. President Mandela planted a tree to mark the occasion, noting:

SANBI artist Gillian Condy painting Mandela's Gold

'I am happiest when I am in the wild because I can listen ... I always feel the force of that sentiment when in this environment. I am very happy that you have done me the honour of being associated with this remarkable place, Kirstenbosch.'
NELSON MANDELA

President Nelson Mandela's visit to Kirstenbosch in 1996

New directions in a changing South Africa

South Africa is home to some 4 000 species of succulent, 40 per cent of the world total. Aloes, mesems, crassulas portulacarias and many other species were planted out in the early years of the Garden's development and have provided a regular attraction to winter visitors throughout the Garden's long and changing history. Here, Aloe striata *is surrounded by the silver-green of* Senecio crassulifolius, *with* Cotyledon orbiculata *and* Aloe ferox *in the background.*

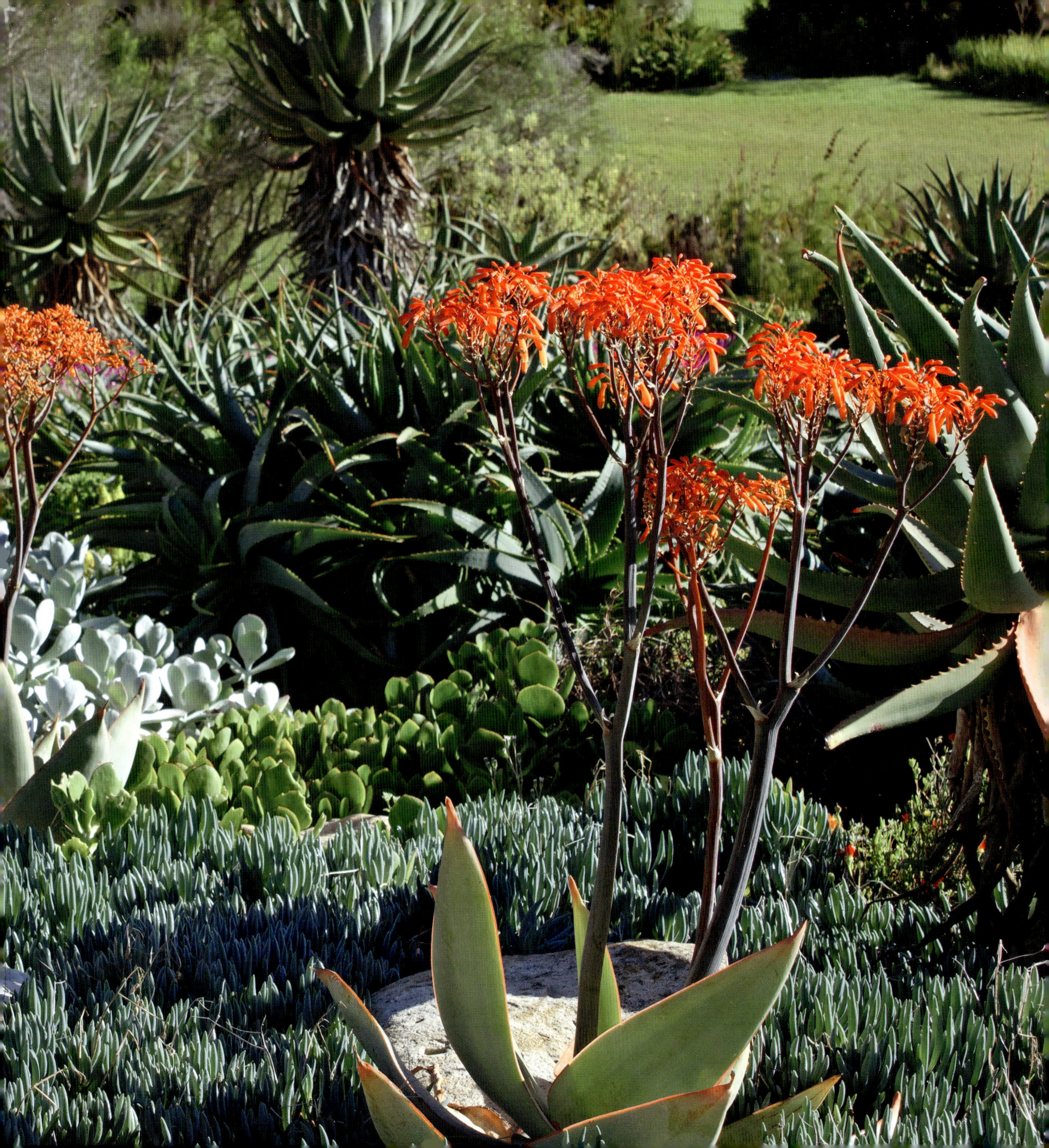

A first challenge for the new CEO was to integrate the talents of the two organisations into a single entity with a common vision and a shared culture. The old north-south tensions remained, but careful consultation, strategic planning sessions, team-building activities and doggedness helped develop a new personality for the organisation. The dramatic political changes of the 1990s provided a dynamic intellectual environment that encouraged positive rather than negative responses. The NBI's goals were ambitious: 'To become a world leader in the development of botanical gardens, in plant conservation and education, and in botanical research by the year 2000; to ensure that its programmes are relevant to South Africa's needs; and to attain excellence in all of its activities'. Despite resistance from some pessimists, these goals were soon embraced by the NBI team.

Unlike the National Herbarium in Pretoria – which had large, modern facilities, a strong body of researchers and a healthy budget – Kirstenbosch, as the new institution's head office, was severely resource limited. Once agreement had been reached on the new organisation's corporate strategy, structure and programmes, it became critical to source finance and develop facilities for the Garden. Government funds were not expected to increase at a pace commensurate with needs – other sources would have to be accessed.

The Kirstenbosch Development Campaign

A new vision and approach was needed. This did not come from the Board, or from hired consultants, or from in-house think-tanks. It came from an old friend and benefactor who had been familiar with the Garden from Pearson's time: Mary Mullins challenged the new management to make it 'The Kew of the southern hemisphere'. For this no new plans were needed, but rather a process that would capture the imagination and support of the vast number of people who loved Kirstenbosch. The mood was right for a Kirstenbosch Development Campaign. An initial target was to raise R20 million by 1995. The story of how R65 million was raised by 2005 is described in chapter 10. Here we summarise the process (see panel alongside).

BUILDING THE INFRASTRUCTURE

The many existing development plans were studied and rationalised according to one principle – to locate all new buildings below the old Rhodes Drive. How to do this was a challenge: the area was already occupied by a jumble of workshops, offices, storerooms, the tattered remains of workers' change rooms and toilets dating from the 1930s, a rather swampy old pasture and an abandoned quarry. A process of musical chairs had to be implemented – juggling old and new buildings and their varied functions. Without a dedicated budget for the long shopping list of needs, an approach of 'strategic opportunism' was adopted. The general plan of needs was already there – now management had to match unpredictable funding sources and unexpected opportunities to fortuitous timing of events.

The first project, in 1991, was to complete the restoration of Pearson House to serve as a modest but comfortable office unit and Board meeting facility. Here the skills of Cape Town architect David van den Heever proved a great success. The NBI's Daan Botha, Director of Gardens and Horticulture, monitored construction works. Botha and Van den Heever made an excellent team for the whole programme. Next came new change rooms, mess rooms and toilets for the Garden workers – completed in 1992 with funds squeezed out of corners of the existing budget.

In 1993, Mary Mullins donated R120 000, which was used to build a 1.2-kilometre walkway through the then inaccessible Protea Garden. In 1994 the first phase of the new Kirstenbosch Research Centre was built on land added in the northeast corner – 'Newlands Heights' – which the government donated to the Garden after the Botanical Society led a massive protest against plans to sell the land to a real estate developer. To secure the deal, R3.2 million had to be raised within three months – ably led by Kay Bergh, Board member, Botanical Society stalwart and the driving force behind the fund-raising campaign that made the whole venture possible.

One of the first projects of the Kirstenbosch Development Campaign was the construction of the Mary Mullins Walkway through the Protea Garden, previously inaccessible to most visitors. In the foreground are orange and yellow forms of Wild Dagga Leonotis leonurus.

The old, stolid 'Bauhaus-design' Environmental Education Centre was modified in 1994 to become a much more functional facility: the 'Gold Fields Environmental Education Centre', named for its corporate sponsor. By 1995, the Department of Environmental Affairs and Tourism had joined the campaign, making R10 million available to complete the research centre. The botanical collections of the Compton Herbarium could now be transferred into the new Kirstenbosch Research Centre (KRC), and the Garden Administration into the old Compton Herbarium building, which was now newly refurbished. The process was working at full speed.

The following year, the project that the Botanical Society had championed for over a decade – a new conservatory (see pages 126–127) – could be built on the site vacated by the various offices

Despite the absence of a budget, the Board approved the restoration of Pearson House as the National Botanical Institute's first Head Office.

New directions in a changing South Africa

The old workshop facilities were replaced in 1997 by the new Visitors' Centre.

In 1990, the mess rooms and toilets for the Garden workers were in a sorry state.

New facilities for the Garden staff were built on the shady Cork Oak lane above the Nursery.

and storerooms of the Garden administration. Here another Cape Town architect who had inspired the entire conservatory idea, Julian Elliot, produced an aesthetically elegant and functionally efficient home for succulents. Collected from throughout southern Africa, these fascinating plants were expertly landscaped and curated by Ernst van Jaarsveld.

By the end of 1996, the basic elements of a modern Botanical Garden were in place – research centre, education centre, administration centre, and a large, elegant conservatory. But much more was needed to reach the goal of becoming 'the Kew of the southern hemisphere'.

Another wave of fund-raising in 1997 won the support of the De Beer's and Anglo American Chairman's Fund, Old Mutual, and government. A large, multifunctional Visitors' Centre could now be built at the restored original entrance to the Garden. At the time, the development of the new facility cost R11.5 million. It comprised a conference centre, garden shop, rest rooms, information kiosks, videorama, coffee shop, and parking for 180 cars and 10 coaches. But the Garden still needed a restaurant to replace the old prefabricated, 'temporary' building erected on the site of the original tea house that had burnt to the ground in 1982. After a prolonged and, at times, emotional debate, and partially funded by government, the new R9-million restaurant was completed in 1998, discreetly positioned in the lower Garden. In both these major projects, the young architect David Lewis brought his innovative skills and sensitivity to the setting.

A Garden without adequate plant-propagation facilities and a living collections repository is not a proper 'botanical' garden. The Kirstenbosch branch of the Botanical Society took on the responsibility of funding a R2.7-million suite of glasshouses, positioned on the warm, sunny, north-facing slope above the conservatory, the site of Pearson's first venture into economic plant production. The Millennium Glasshouses are now the gene bank of the Garden's rich diversity of bulbs, succulents and other plants of special conservation importance.

Promoting the horticultural value of South Africa's indigenous plants has always been a central objective of the institute, and Kirstenbosch has supplied surplus seeds and plants and provided gardening advice to the public throughout its history. It was logical that a centre for home gardening should be included in the Garden's facilities. With sponsorship from Pam Golding Properties, and additional funding from government and the Botanical Society, the penultimate major project was completed in 2003: a garden centre, entrance gate and tea house, plus upgrading of the old parking area and preparing an extensive marquee lawn on the site of the house built in 1812 by Henry Alexander, then Colonial Secretary.

The final project in the major infrastructure development programme was a consequence of the successful partnership model that NBI had followed in establishing itself as a lead player in South Africa's biodiversity sector. The Kirstenbosch Research Centre (KRC), with its Compton Herbarium, Molteno Library,

Allan Bird Ecology Laboratory, Leslie Hill Molecular Laboratory, plus conference and workshop facilities, had attracted a vibrant network of conservation scientists from many governmental and non-governmental agencies. Kirstenbosch had already become a hub of new co-operative programmes, with the secretariats of these organisations housed in various corners of the Garden. To achieve the maximum 'critical mass' of this energy, a proposal was made to the Rufford Maurice Laing Foundation, a charitable trust in the United Kingdom, for a grant to build a Centre for Biodiversity Conservation adjoining the KRC. The proposal was strongly supported, on the understanding that the facility would be reserved for NBI's partner NGOs and collaborative projects based at Kirstenbosch. A grant of R12 million was made and the new centre opened on 1 September 2005.

SANBI

The multifunctional Visitors' Centre, Botanical Society Conservatory, Millennium Glasshouses, workshops and old Compton Herbarium fill the land between the new Rhodes Drive and the historic avenue of Morton Bay Figs and Camphor Trees.

New directions in a changing South Africa   105

The Kirstenbosch Development Campaign was a resounding success. The approach of 'strategic opportunism' resulted in the full diversity of the Garden's infrastructural needs being met through a combination of individual private donations, corporate sponsorships and government grants. Of the R62.7 million invested between 1992 and 2005, R33.7 million (53 per cent) came from donations. The power of the partnerships created through the campaign had lasting results (as we will examine in chapter 10). From being 86 per cent dependent on government grants for the operations of the Garden when Huntley took over management of the NBI in 1990, Kirstenbosch Garden was operating at a profit by 2006, when he stepped down from the position of CEO in order to facilitate transformation in the Institute.

But resources and skills were not concentrated on buildings alone. Kirstenbosch's unique 'competitive advantage' over other gardens around the globe is the combination of horticultural talent with the wide palette of colour, form and long-flowering behaviour of our flora. Unlike most 'northern' plants, our proteas, ericas, agapanthus, aloes, etc, ensure a continuous display of colour all year. During the 1980s and 1990s, the Kirstenbosch team had, under the leadership of Curator John Winter, greatly expanded the range and display of species and cultivars. By the end of the millennium, Kirstenbosch had new plantings of a wide diversity of restios – new to horticulture since the discovery of their germination cues. Sweeps of many *Plectranthus* species added colour to the tones of green in the Dell, while landscaping around new buildings both softened and enhanced their impact.

From NBI to SANBI

Initially a flora- and vegetation-focused organisation, the NBI took advantage of the changing field of conservation science – and the opportunities that arose with major funding from the Global Environment Facility – to develop large, multidisciplinary, longer-term programmes. From early in the 1990s it had championed strong partnerships and adopted a 'managed network' business model. With the increasing flow of investment into democratic South Africa and many new opportunities being offered – and given NBI's record of good corporate management – NBI was becoming the preferred partner to head up large and complex programmes.

It provided co-ordination, management and 'honest broker' support. By the end of the decade, it was administering over R150 million in donor funds for projects such as the Cape Action for People and the Environment (CAPE), the Succulent Karoo Ecosystem Programme (SKEP), the Global Invasive Species Programme (GISP) and the incipient Grasslands Biome Project.

NBI also explored new approaches to bioregional programmes, leading discussion on new concepts such as 'mainstreaming' and 'sustainable landscapes'. The Centre for Biodiversity Conservation at Kirstenbosch housed the secretariats of CAPE, SKEP, GISP, the WWF Table Mountain Fund, Conservation International's Southern African Hotspots Programme, and IUCN's Species Survival Commission. By 2005 it had 11 partner organisations working out of the shared facilities in Kirstenbosch, and many more organisations regularly using its conference rooms for scientific meetings, conservation planning workshops and to celebrate project successes.

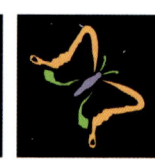

106 KIRSTENBOSCH

From botany to biodiversity – the establishment of SANBI

In 2002, 10 years after the United Nations Conference on Environment and Development (UNCED) in Rio de Janeiro, the follow-up 'Rio+10' Conference was held in Johannesburg – the World Summit on Sustainable Development (WSSD). Rio had been a watershed meeting for biodiversity, resulting not only in the establishment of the Convention on Biological Diversity (CBD), but also in international conventions and agreements on climate change, desertification and biosafety. Unlike many similar conferences, UNCED led to the creation of a mega-fund for action – the Global Environment Facility (GEF).

These initiatives had profound impacts on how the world community shaped environmental policy and practice. NBI, in developing its Corporate Strategic Plan for 1994 to 2000, framed its new research programmes around the anticipated requirements of government in respect of biodiversity, climate change and desertification. Similarly, the South African government responded energetically by formulating its post-1994 law-reform process to harmonise with international trends. The National Environmental Management Act of 1998 was strongly influenced by the Rio outcomes, and the WSSD experience helped mould the subsidiary National Environmental Management: Biodiversity Act (NEMBA) No. 10 of 2004, which established the South African National Biodiversity Institute (SANBI).

The formation of SANBI came as a natural response to the widening perspectives of NBI. As a result of the experience and credibility gained at Kirstenbosch through the 1990s, SANBI could easily adapt to the new and expanded mandate of the Biodiversity Act. In the years since SANBI's establishment on 1 September 2004, and the relocation of its head office from Kirstenbosch to Pretoria in 2008, it has responded to its new responsibilities efficiently and effectively. Kirstenbosch is now but one of SANBI's many assets.

From its humble and difficult beginnings in 1913, the Garden has weathered many storms and witnessed many changes in corporate identity and leadership. But it has always been the flagship of the organisation, the single entity that is both an unmistakable brand and an unmatched icon for South Africa's floral heritage.

What's in a logo?

Institutions have, from the earliest times, identified themselves with insignias of varying form. Kirstenbosch, within the organisational framework of the National Botanic Gardens, National Botanical Institute, and finally the South African National Biodiversity Institute, has used a range of logos. The choice of logos used by Kirstenbosch and the Botanical Society reflects changing corporate identities.

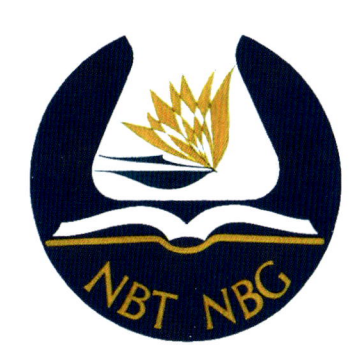

New directions in a changing South Africa

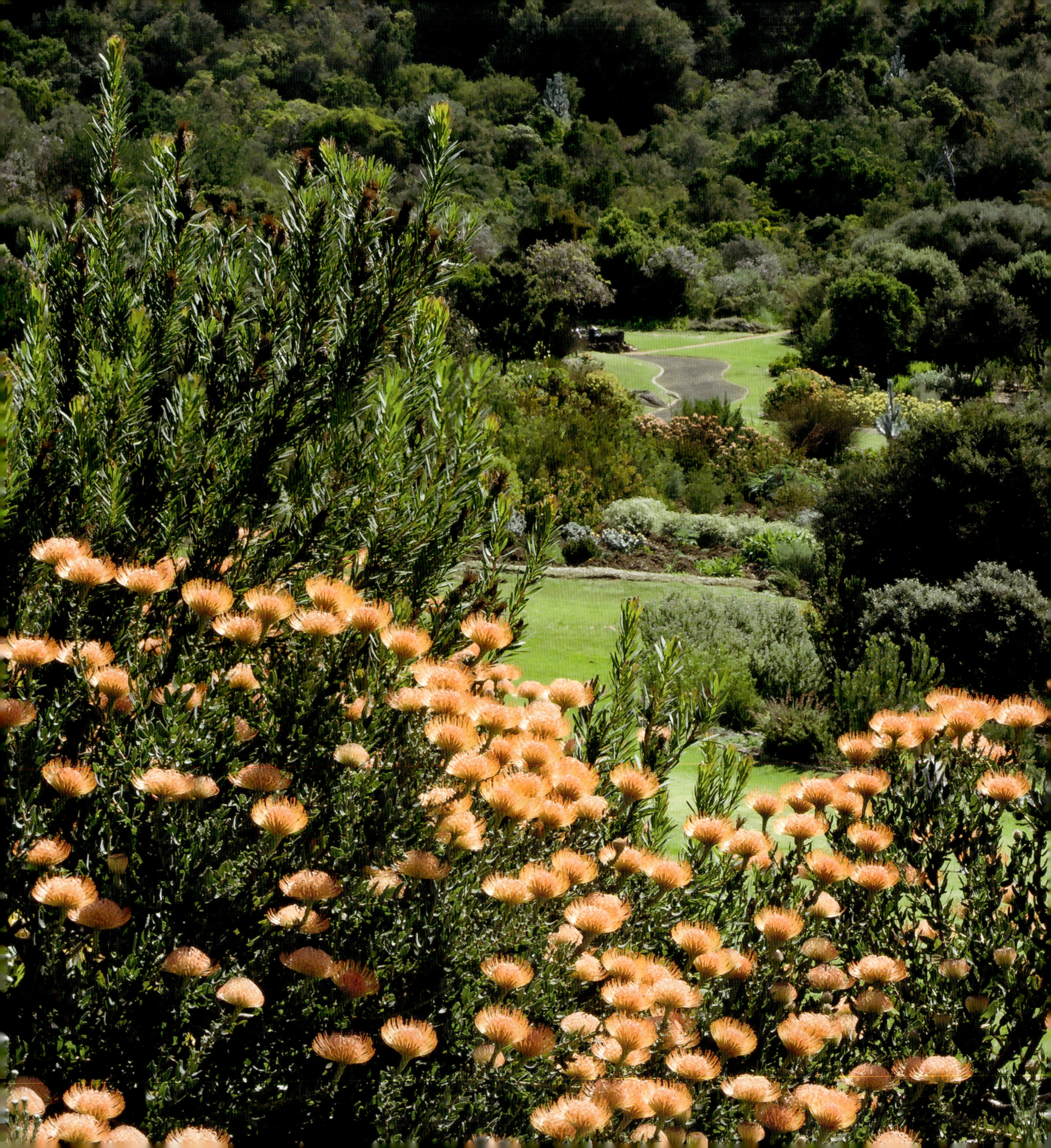

CHAPTER SIX

A Garden for all seasons

Gardens within the Garden

The objects immediately surrounding us were purely sylvan. Here I beheld [Nature's] perfection in the sweet harmony of soft colours and tints of every gradation, speaking a language which all may understand, transfusing into the soul a delight which all may enjoy, and which never fails, at least for the time, to smother every uneasy sensation of the mind.

WILLIAM BURCHELL
Travels in the Interior of South Africa, *1822, nearly a century before the Garden's foundation, and some two centuries before it reached its present state of development*

The heart of the Garden presents soft lines, rich colours and the evergreen verdure of fynbos merging into the forests of the mountain. The striking pincushion in the foreground is Leucospermum cordifolium x tottum *'Caroline'.*

The Overberg Pincushion Leucospermum oleifolium *fronts the Garden's sweeping, rugged mountain backdrop.*

Adam Harrower

In the shadow of Table Mountain

The first garden established at the Cape in 1652 was situated close to the stream that flows off the north-facing slopes of Table Mountain. We have already referred to the Company's Garden in chapter 1. It succeeded as a source of vegetables for the growing colonial administration, but never developed into a true 'botanic' garden. The reason is simple – it was built on the wrong side of the mountain.

The choice of Kirstenbosch was motivated by the sheer grandeur of the site. There is no evidence to suggest that Harold Pearson did a careful site feasibility study, based on the multiple criteria that are now considered before proposing the establishment of a new botanical garden. He did not consider geology, soils, water, habitat diversity, ease of access or 'stakeholder' opinion. He simply knew that 'this is the place'.

How fortuitous that his choice fulfilled all the key elements critical to the success of a modern botanical garden – a century ahead of his time. Here, in the shadow of Table Mountain, lying on shallow, rough, acidic sandstones, deep granitic clays and peaty alluvial soils, with three permanent streams fed by heavy winter rains, was a site *par excellence* for a garden. On east- and northeast-facing slopes, drenched by the morning sun rising over the distant Hottentots Holland Mountains, the site cools off early in the afternoon as the sun disappears behind its mountainous 'garden wall'.

Seven hundred million years in the making

The most striking feature of Kirstenbosch is the sweeping, rugged mountain backdrop to the soft verdure of the Garden. The geological history of the site is long, complex and all but hidden to the visitor by a deep layer of rocky debris that covers the lower slopes of the natural basin in which it is sited. Only a simplified summary of the Garden's 700-million-year evolution can be given here. The Garden's oldest rocks are from the Malmesbury Group. These are metamorphic rocks formed by the compression of sediments deposited in deep marine basins. They were formed between 700 and 560 million years ago.

Tens of millions of years later – about 540 million years ago – violent disturbances at great depths caused the rocks of the Earth's mantle to melt into a viscous magma, which intruded into the overlying Malmesbury and other systems to cool, solidify and crystallise at depths of several kilometres. This process formed the Cape Granites, large exposures of which are found around the Cape Peninsula, most prominently at the Boulders penguin colony. In Kirstenbosch, granite soils support the Silver Tree grove near Rycroft Gate.

The most prominent of the geological formations to be found in the Garden are the sedimentary rocks of the Table Mountain Group. The sediments that formed these mountains are thought to have accumulated, from some 520 million years ago, along the coastline of the ancient supercontinent called Gondwana.

The Table Mountain Group is represented on the Cape Peninsula by two formations – Graafwater and Peninsula. In Kirstenbosch the Graafwater Formation can be observed along the upper contour path, where its soft, red sandstone sediments have been eroded in some places to form small caverns. The Peninsula Formation is a much thicker, coarser layer of sediments resistant to erosion, and results in the impressive precipices below Fernwood Buttress and the rugged backdrop of Castle Rock.

About 250 million years ago a mountain-building period preceded the breaking up of Gondwana. It is believed that a small oceanic continent drifted into Gondwana, pushing up the deep sediments of the Table Mountain Group to form the Cape Fold Mountains. Erosion followed the mountain-building period, and the thick cap of sandstone over Table Mountain was reduced to half its original height. Even greater erosion events stripped away the landscape between the Peninsula and the other Cape mountains, creating the Cape Flats.

The recent history of the Cape Peninsula saw the sea level rise and fall during the Pleistocene Ice Age of the last two million years. Fluctuations of over 130 metres in sea level resulted in the exposure of a much wider coastline around the southern Cape, with the Agulhas plain extending over 150 kilometres further south than it does today. Conversely, sea level rise during the warmer cycles of the Ice Age caused the Cape Flats to be flooded, creating an island of the Cape Peninsula, and isolating Kirstenbosch and the whole of Table Mountain from the continent.

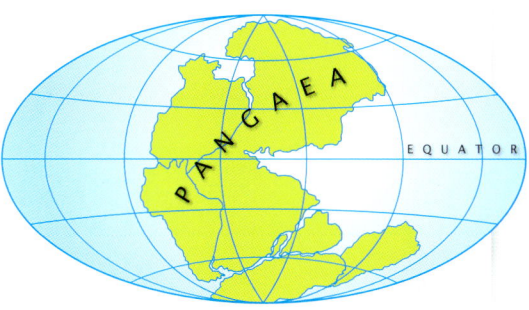

PERMIAN
225 Million years ago

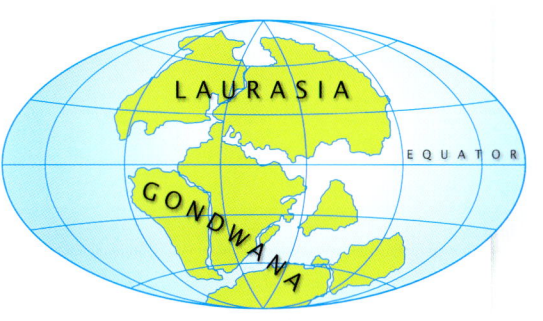

JURASSIC
135 Million years ago

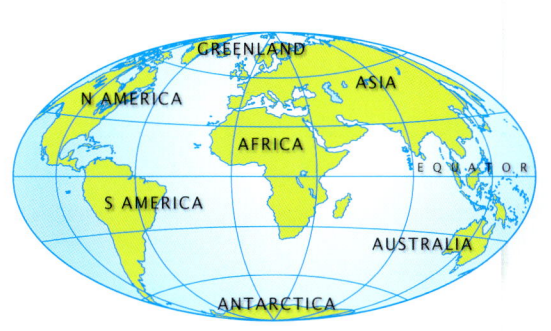

PRESENT DAY

Since the rugged sandstones of Table Mountain were laid down 540 million years ago, the Earth's continents have drifted across the globe. The ancient Pangaea split into Laurasia and Gondwana, and Gondwana, in turn, separated into Africa, South America, Australasia and Antarctica.

A Garden for all seasons   111

A Mediterranean climate

The Kirstenbosch climate is typical of the Mediterranean regions of California, Chile, southern Australia and the Mediterranean basin – warm, dry summers and cool, wet winters. The high, east-facing ridge rising above Kirstenbosch accentuates the temperature and rainfall gradients of the Garden's climate, giving it a unique set of microclimates.

But it has not always been so. The process of continental drift had dramatic consequences for climates across Africa. After breaking away from South America 130 million years ago, Africa drifted northwards: today, the Cape is 20 degrees of latitude closer to the equator than at the time of separation. Some 12 million years ago, the glaciated Antarctic finally broke away from South America, giving rise to the circumpolar ocean and the cold, upwelling Benguela Current along the west coast, stabilising weather patterns in the region.

A stationary subtropical high-pressure cell hovers over the South Atlantic. During summer, low-pressure cells pass eastwards to the south of the Cape, while in winter, the low-pressure cells shift northwards by five degrees of latitude, and so pass over the Cape. As the low-pressure cells develop and approach the Cape, converging masses of cold air from behind the front mix with the warmer air preceding it, resulting in condensation and belts of clouds, and forming the regular cold fronts (about 20 per year) that bring the Cape its stormy, wet winter weather.

It is usual to have delicate high cirrus clouds – often as lacy 'mackerel skies' – preceding the fronts. And, before this, the passage of hot, dry 'berg' winds rushing off the interior plateau is a frequent phenomenon.

A point of enduring curiosity is why Kirstenbosch receives over 1 300 millimetres rainfall per annum, when the city centre, just 10 kilometres distant, receives only 520 millimetres. One might expect that the strong westerlies and northwesterlies of the cold fronts would drop their heavy load of rain on central Cape Town as they reach the steep wall of the Table Mountain complex, which faces the oncoming storms. But the ridge of mountains is too low and too narrow to stop the rain-bearing clouds, which are forced upward and, rapidly condensing, precipitate on the opposite side – precisely above Kirstenbosch. As a front approaches, clouds cascade through Constantia Nek and over the gap between Devil's Peak and Fernwood Buttress, followed by strong winds and pouring rain.

On some occasions the winter gales wreak havoc – in September 2001 the metal roof of the recently completed Visitors' Centre was ripped off. Fortunately the storm struck in the middle of the night, with no people on the site – had it happened a few hours later there might have been fatal consequences.

Whereas winter storms can rage over the Cape Peninsula for several days with each passing cold front, the grey, cold weather is often broken by gloriously sunny, champagne days in mid-winter. A contrasting situation arises during spells of hot summer weather. The dominance of strong southeasterly winds in

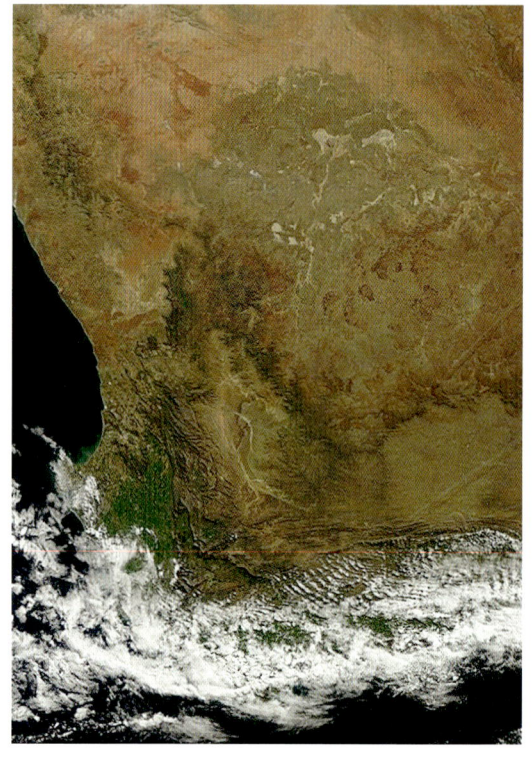

As a weak cold front approaches the Cape, a low-pressure system along the coast draws hot air off the interior, resulting in a 'berg' wind, with temperatures at Cape Town surging to over 30°C, before dropping by more than 15° as the front passes by. (Satellite images courtesy of the South African Weather Services)

A display of particularly forbidding mammatocumulus clouds looms over the Cape as a cold front passes by.

A winter northwester heralds the passage of a cold front, with clouds pouring through the gap between Fernwood Buttress and Devil's Peak.

Condensation of fog on restios such as Ischyrolepis subverticillatus is a key source of water on Cape mountains during the summer.

A Garden for all seasons

Four seasons in one day

Capetonians are given to saying that, if you don't like the weather, wait around for a few hours. The weather station at Cape Town International Airport weather station has recorded changes of over 20°C in just a few hours, especially as hot 'berg' winds from the interior arrive on cool winter mornings. This area was aptly named the 'Cape of Storms' by Portuguese navigator Bartholomeu Diaz some 500 years ago, and the dynamic nature of the area's weather often produces four seasons in one day.

summer forces moist air up and over Table Mountain. A band of stratus cloud forms the famous 'tablecloth' that pours over the lip of the Table, evaporating as it mixes with the warm dry air rising off the basin in which the city of Cape Town lies. Kirstenbosch is largely protected from the onslaught of the strong southeasterly summer winds – an island of calm within the storm – but there is enough of a breeze to bring relief on hot days. On the summit, the condensation of summer clouds on the surfaces of plants and rocks, when added to the winter rainfall, effectively doubles the annual deposit of moisture – a phenomenon first accurately measured in the 1890s by that pioneer of fynbos ecology, Rudolf Marloth.

At the popular Kirstenbosch Summer Sunset concerts, visitors in light summer dress soon find themselves searching for a fleece as the afternoon sun disappears behind Maclear's Buttress and cold air rolls down the mountain slope. This cool, moist, microclimate provides the required conditions (as well as protection from fire) for the indigenous Afrotemperate Forests that clothe the ravines above the Garden, and permits the planting of a wide diversity of species from the higher reaches of the Cape Fold Mountains – home to the richest diversity of the Cape flora.

Characteristic, too, of the Kirstenbosch climate is the mist and fog shrouding the Garden's lower reaches on many autumn mornings. Early visitors climbing up Nursery Ravine are delighted to emerge from the foggy cover of the lower Garden slopes into the bright morning sunlight, with the Cape Flats still hidden below the mists.

Delicate high cirrus clouds announce the arrival of a cold front on an early spring day. Seen flowering are Bulbinella latifolia *(yellow),* Lampranthus spectabilis, Lampranthus amoenus *(pink) and* Drosanthemum speciosa *(orange).*

On a foggy autumn morning, the lower garden is covered by a damp shroud while the upper garden is bathed in sunlight.

The Cape Flats lies hidden below a sea of cloud embracing the mountain's lower slopes. The Garden's 150-million-litre reservoir is in the foreground.

A Garden for all seasons

Designing Eden

Botanical gardens, as national or private institutions, went through a vibrant renaissance during the late 20th century, and this positive turn of fortune continues to this day. Many new and exciting gardens have been built in the space of a few years, often at huge expense and following ambitious master plans. The Eden Project in Cornwall is a fine example of such initiatives, as are new gardens in Wales, Singapore, Oman and Australia. Garden design has become a major industry, and landscape architecture a respected profession. But Kirstenbosch was different – it has never had a formal horticultural 'master plan'. Compton remarked:

> *'… landscaping at Kirstenbosch was always rendered futile by the grandeur and diversity of its setting – the landscape was already there … in the special conditions of Kirstenbosch, a fixed design too rapidly carried out might have had worse results'.*

Until the defining work of Lancelot 'Capability' Brown in the 18th century, garden design in Europe was founded on the assumption that man should impose his vision on the site – not that the site should suggest the design. Brown took a different approach, looking first at what the landscape was capable of providing. This thinking was clearly at the heart of what Mathews and his successors created at Kirstenbosch.

The terrain and its habitats determined the early design of the Garden: the moist dell and the waterlogged 'bog garden', the dry 'aloe koppie', the well-drained protea section. These features, along with the extensive undulating lawns, clumps of trees and shrubs, discreet water features, rough-cut walls using local stone, and integrated plantings of selected fynbos and other floristic groups, give the Garden its unique impact. The mountain backdrop has afforded the blending of intensively landscaped gardens across the lower reaches of Kirstenbosch with a gradual transition into the natural vegetation along the upper slopes.

Some attempts were made to group plants on a systematic basis, according to their families, in the tradition of medicinal gardens, but this was short-lived. So, too, was the attempt to grow a Cape Chestnut avenue, described by Compton as 'leading from nowhere to nowhere'. An orchid garden, in honour of the great benefactor and amateur botanist, Harry Bolus, had to be abandoned after several years of frustration.

At first, Mathews had to make do with the normal variations and uncertainties of rainfall. But as the Garden grew in size and complexity, more reliable water resources were essential. A small reservoir and boreholes were replaced in 1988 by a large reservoir with a capacity of 150 million litres. In 1990 the Botanical Society provided funding for an extensive automated irrigation system, sufficient to keep the Garden's 12 hectares of lawns lush and green through the hot, dry summer. Old metal pipelines, corroded by the acid waters of the mountain streams, were replaced with resistant PVC pipes. Automated pumps could ensure irrigation at night to preclude the sudden saturation of strolling visitors. Equally important, in the 1980s, a large 'compost factory' was developed, using all garden cuttings and clippings to produce 1 500 tonnes of rich, organic compost each year.

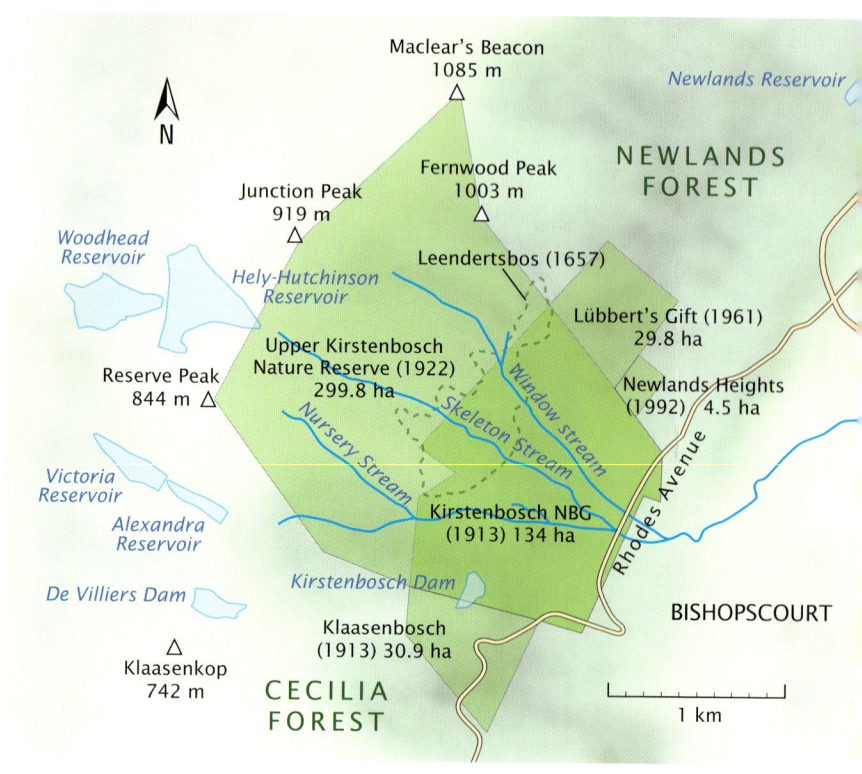

At establishment on 1 July 1913, Kirstenbosch covered 134 hectares. Gradual additions as indicated on this map brought the total area owned by the Garden to 199.2 hectares by 1992, with an additional 299.8 hectares being managed as part of the Garden since 1922. The position of the original Leendertsbos is indicated by a dashed line.

Dedicated Kirstenbosch families

Kirstenbosch not only grows plants – it also grows generations of loyal staff members. Consider master stonemason Abraham 'Awie' Basson, who started working at Kirstenbosch in 1926, at 12 shillings a week. His father had preceded him at Kirstenbosch, as had his mother. Retiring at 65, after 48 years in the Garden, he recalled his life at Kirstenbosch as 'working hard, but working jolly'.

The commitment of the Kirstenbosch team is no better exemplified than by the 52 years' service of David 'Driver' Mclean. At the age of 13 he joined the staff as a messenger – the bicycle provided was far larger than David. After serving in the S.A. Army in the Middle East from 1939 to 1943, he returned to Kirstenbosch as a driver of tractors, trucks and buses, as well as his beloved front-end loader.

The Jacobs family is another of the many that have been associated with the Garden over several generations. Dennis Jacobs, born in 1928, joined Kirstenbosch in 1963 and retired in 1993. Three of his sons, Andrew, Clive and Freddie, continue their careers in the information, nursery and administration sections of the Garden.

The exemplary service of the Kirstenbosch team is reflected in the 'class of 1973' which had eight long-service awardees, with a total of 310 years of service between them!

David 'Driver' Mclean retired after 52 years at Kirstenbosch – seen here with Brian Rycroft, Director from 1953 to 1983.

The Jacobs family are Kirstenbosch stalwarts: from left to right – Andrew, father Dennis, Clive and Freddie.

Today, like few other gardens, Kirstenbosch nestles within a vast protected area – the Table Mountain National Park, proclaimed in 1998. Together, they form part of the Cape Floristic Region World Heritage Site, inscribed by UNESCO's World Heritage Commission in 2004. Extreme good fortune accounts for this situation – a sympathetic government and a generous public.

Kirstenbosch consisted of 134 hectares at establishment on 1 July 1913. Later that year, 30.9 hectares of the Klaasenbosch Estate were transferred to the Garden. In 1922, 299.8 hectares of mountain catchment (the Upper Kirstenbosch Nature Reserve) came under the Garden's control. This land, reaching Maclear's Beacon, the highest point on Table Mountain, includes the catchment of the three perennial streams that are the life-blood of Kirstenbosch. In 1961 a gift of 29.8 hectares was made to the Garden by Mrs I.G. Lubbert, while 4.5 hectares (Newland's Heights) were added in 1992. Today Kirstenbosch comprises 199.2 hectares. The 299.8-hectare Upper Kirstenbosch Nature Reserve was transferred to the Table Mountain National Park in 1999, but continues to be managed as an integral part of the Garden.

Construction of the Garden's 150-million-litre reservoir in the 1980s ensured an adequate supply of water for the gravity-fed irrigation system.

A Garden for all seasons

The rise of indigenous plant horticulture

The famously rich South African flora presented an exciting challenge to the Garden's horticulturists. The talent, patience and skills (and no doubt the Kew training) of many of Kirstenbosch's horticulturists – Mathews, Thorns, Werner, Winter – changed the face of indigenous plant horticulture in South Africa. Soon after the establishment of Kirstenbosch, plants began to be donated to the Garden from all corners of the country, supported by free rail transport in the early years, providing an almost limitless supply of new and exciting material to test and introduce into the Garden. Mathews built modest greenhouses to propagate the vast collection of cuttings and seeds that were flowing in – over 2 000 accessions arrived in 1915 alone – and the pace did not slow until the 1930s. Under the leadership of Compton, extensive field collecting trips were made, filling gaps in the living collections, and expanding the herbarium.

Kirstenbosch is deservedly famous for its splendid displays of pincushion proteas such as the yellow Leucospermun reflexum var. luteum *(front) and* L. cordifolium *(behind).*

Stone Pines, planted in the 19th century, loom below Castle Rock.

A tolerant approach

In spite of the sensitive Garden design, at the time of its establishment in 1913, years of neglect had resulted in a dense cover of alien species – oaks, pines, poplars, gums and brambles. During the Garden's early years, acorns provided a necessary source of income (as pig-fodder), and the few surviving old oaks, along with scattered Stone Pines (some planted in the mid-19th century) were retained – and form part of the 'cultural landscape' to this day, offering welcome shade. Resistance to the removal of the oaks was passionate, and so a policy of gradually removing old, weak trees has been followed; most have, over the years, been replaced by indigenous species.

On the other hand, an energetic campaign to remove all invasive alien species has been very successful. This, together with a policy of fire prevention, has allowed the natural forest and fynbos to re-establish across the mountain slopes and ravines.

Gardens within the Garden

Given the richness of the Cape Floral Kingdom, emphasis at Kirstenbosch has always been on the fynbos, which is characterised by three families – Proteaceae, Ericaceae and Restionaceae, plus a rich diversity of bulb species. Fynbos, or 'fine bush', refers to the finely leafed plants, typified by the many *Erica* species, but also a characteristic found in many other Cape plant families. Fynbos vegetation is prolific in species, but some are difficult to cultivate, as early attempts at Kirstenbosch were to reveal.

Proteas have a relatively long history of successful cultivation, even though they fall victim to the root fungus *Phytophthora cinnamomi*, making them prone to sudden death. By 1800, the gardens at Kew had 20 species in cultivation, and the royal gardens in Vienna and Paris were soon to follow. South Africa has 387 species of *Protea*, with 220 species in the Kirstenbosch living collections, displayed in an extensive Protea Garden accessed by a 1.2-kilometre walkway. The King Protea *Protea cynaroides* represents the country in a diversity of emblems – national flower, national cricket team and numerous brands.

Sadly, the once abundant Silver Tree *Leucadendron argenteum*, which so impressed Burchell and other early visitors to Kirstenbosch, and which is endemic to Table Mountain, has been dramatically reduced in numbers by

Clearing and developing the Erica and Protea gardens was a major project of the 1970s.

A Garden for all seasons

the root-rot fungus *Phytophthora*. The spores of this fungus quickly disperse through irrigation water, invading a plant's vascular system and literally cutting off its water supply. Within days of infection, the leaves of a large Silver Tree turn gold, then brown, before the whole tree dies. The upper lawns of the Garden once carried a woodland of elegant Silver Trees – many up to 10 metres in height – of which only a few now remain.

The more humble Ericaceae – better known to the British visitor from the heathers of Scotland than the mountains of South Africa – has no fewer than 682 species occurring in the Fynbos Biome and 104 on the Cape Peninsula alone. Kirstenbosch has 252 species of this rather difficult garden genus. Like proteas, ericas are very vulnerable to fungal attack.

In addition, many *Erica* species are narrow endemics, limited in their natural distribution – in many cases to a single mountain or coastal plain. Many of these endemics are threatened by land transformation such as agriculture, urbanisation or the invasion of alien vegetation. Over 90 per cent of the landscape of many lowland fynbos vegetation types has been completely transformed over the past three centuries. Some 126 *Erica* species in the Kirstenbosch collection are threatened in the wild, while two species, *Erica verticillata* and *E. turgida*, are extinct in the wild (see page 140).

The Erica Garden was started in 1972 by John Winter (who would become Curator from 1979 to 1999) with the assistance of Dickie Bowler – a member of the Bowler family, which has been associated with the Garden since its founding. A well-drained site was chosen, on acid quartzitic soils on the upper slopes of the developed Garden, and large boulders were positioned to enhance the visual impact of the beds and to create microhabitats for the different species. Each of nine beds, separated by rolling lawns, represents a particular phytogeographic region within the Cape Floral

Special selections of the Forest Lily Clivia miniata *are being trialled in the Enchanted Forest.*

False Buchu Agathosma ovata *'Kluitjieskraal' makes an attractive mauve edging to plantings of shrubby ericas. The buchu family (Rutaceae) has many species that are of value in traditional medicine.*

Kingdom, presenting the endemic and rare species unique to it. Ericas from the different regions grow together with their associated fynbos species, including selections from the protea, restio, buchu and fynbos bulb families.

The aromatic family Rutaceae, or buchu, well known for its use in traditional medicines and as an essence for perfumes, is also rich in species: 314 in South Africa, with 55 of these represented in the Garden. It is the Rutaceae that give the fynbos its characteristic aroma – overly pungent to some but loved by many. Crush the leaves of any rutaceous species and enjoy the rich, musky fragrance.

Until recently, the Restionaceae – or Cape reeds – were almost unknown to gardeners in South Africa and beyond. With 314 species in the country, few were cultivated in the Garden until some innovative field research revealed the germination cue for restios (see page 176). This led to the development of a new Restio Garden, with trial beds of 55 species. The majority proved to be excellent structure plants for landscaping, especially in large gardens and estates.

Kirstenbosch offers shaded habitats for many forest-floor species, including 40 species of *Streptocarpus*, although it is mostly a summer-rainfall group and suffers from the hot, dry days of the Cape summer. Similarly, the genus *Clivia*, or Forest Lily, comes from South Africa's summer-rainfall area. But John Winter has built up a valuable living collection of the genus, including the first breeding stock of the rare, winter-rainfall, and aptly named *Clivia mirabilis*. This species was discovered only in 2000 in a shaded but dry valley near Nieuwoudtville, in the winter-rainfall region of the Northern Cape.

Pioneer species such as the Keurboom, Tree Fuchsia, Turkey Berry, Wild Pear and Cape Myrtle have been used as nursery plants to provide shade in the 'Enchanted Forest', a relatively new development on the warmer, north-facing slopes of the Garden, where more than 450 different trees (representative of some of South Africa's over 1 000 species of tree flora), have subsequently been introduced. John Winter initiated this project in 1980, and the Forest's dense, shady habitat created has been used by Ernst van Jaarsveld to introduce the many species of *Plectranthus* he has collected in the forests of KwaZulu-Natal and the Eastern Cape. Wide sweeps of pink, mauve, blue and white plectranthus now cover the forest floor in autumn, with red and orange clivia appearing through late winter and early spring, and clumps of streptocarpus dotted through the deeper shade.

A Garden for all seasons

Many new selections of horticultural interest have been introduced to Kirstenbosch by the Garden's plant hunters. In particular, the genus Plectranthus *has offered many exciting finds. These include the white-flowering* Plectranthus ecklonii *'Tommy', the purple* P. ecklonii *'Medley Wood' and the pink* P. ecklonii *'Erma'. A spreading crown of* Schotia brachypetala *provides an ideal habitat for these shade-loving species of the coastal forests of eastern South Africa.*

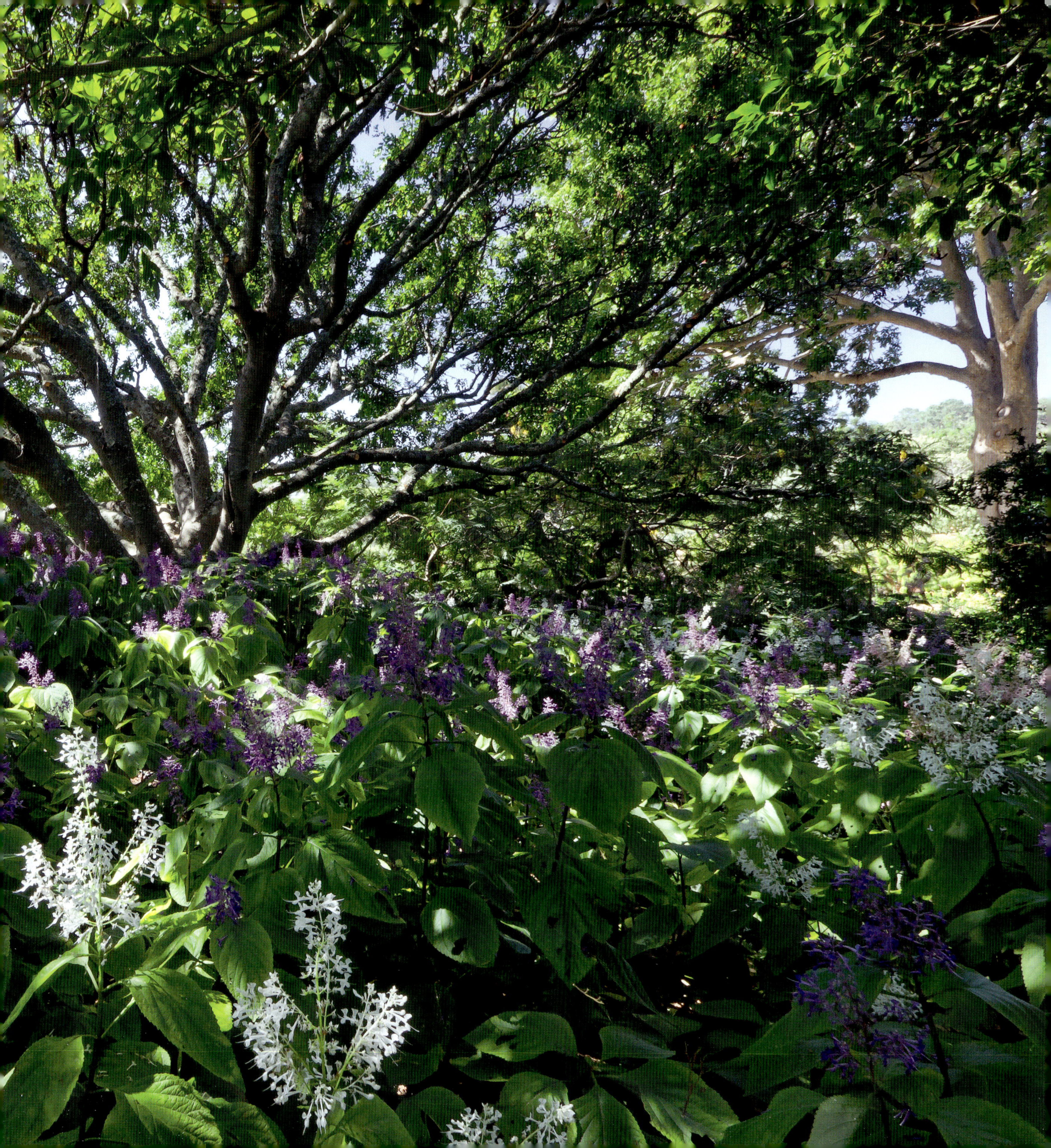

A horticultural tightrope

Water in the right quantity and at the right time, and regular composting of the lawns and beds, are as critical to the success of the Garden as is its special setting. But both watering and composting carry dangers.

Fynbos species are adapted to nutrient-poor soils and dry summers, while the lush, thirsty kikuyu and buffalo grass lawns that intersperse the flower beds need regular watering and feeding. Too much watering in summer promotes fungal growth in sensitive fynbos species – in particular, the root-rot fungus *Phytophthora* – which is often fatal. The absence of fires in Kirstenbosch, which are the natural agents of sterilisation in fynbos ecosystems, puts the fynbos at even greater risk.

What is more, fynbos species dislike having their roots disturbed, making it difficult for gardeners to control the weeds that grow prolifically in summer, and which further deplete the poor soils as well as make the Garden look unsightly.

To accommodate these preferences, the fynbos beds are regularly covered with a dense layer of wood chippings, serving the dual role of smothering weed growth and retaining moisture. Compost has also been used as a mulch in the erica and some protea beds, giving the added advantage of providing regular, but low, levels of nutrients. Fynbos horticulture is not for the faint-hearted!

The first sign of infection by root-rot fungus in a Silver Tree is the rapid wilting of terminal leaves, followed by a golden glow, and death within a few weeks.

The Garden's compost factory supplies the fynbos beds with a dense cover of wood chippings to conserve water and suppress weed growth.

Mass displays of mesems (locally known as 'vygies') include purple Lampranthus amoenus *in the foreground, with red* Drosanthemum speciosum *and yellow* Drosanthemum bicolor *higher up the slope.*

Adam Harrower

Creating a haven for succulents

South Africa, at last count, had some 4 000 species of succulent plants – about 40 per cent of the global total. They range in size from the tiny stone plants of the Knersvlakte along the west coast to the giant baobabs of the bushveld. Usually found in hot, arid climates, succulents store water in their leaves, stems or roots to tide them over extended dry periods. They have one feature in common: a dislike of too much rain. Kirstenbosch is very wet by comparison with the habitats of most succulents, particularly in winter. The Garden's first Curator, Joseph Mathews, used innovation and a careful choice of species to present a grand display of South African succulents in the rock garden that commemorates his work. By transporting many tonnes of huge sandstone and granite boulders and wagonloads of gravels to a well-drained, sunny slope, he created an environment in which an impressive collection of large tree euphorbias, aloes, sansevierias, portulacarias, mesembryanthemums and other succulents could thrive, despite the Garden's high rainfall.

The Mathews Rockery has served the Garden and its visitors well for nearly a century, but can accommodate only a limited range of species. What the Garden needed was a large, airy, dry and warm conservatory, in which a representative collection of the country's succulents could be displayed. This need became a major, almost obsessive fund-raising project for the Botanical Society, and focused the energies and passions of the Kirstenbosch branch of the society from the late 1980s until the Conservatory was opened in 1997. Today the succulent collection exceeds 700 species – including specimens from the Mesembryanthemaceae, Crassulaceae, Euphorbiaceae, Apocynaceae, Portulacaceae, Aloaceae and Vitaceae families. While most succulent species at Kirstenbosch are now found in the Conservatory, mass displays of mesems in bloom line the main walkway into the Garden, while visitors are treated to the exuberant flowering of aloes in the Mathews Rockery through winter.

A Garden for all seasons

The Botanical Society Conservatory

To meet the need for a 'desert house', Cape Town architect Julian Elliot designed an innovative, elegant and efficiently cross-ventilated building of 3 000 square metres, sufficient to house a baobab as its centrepiece and a vast collection of succulents. Horticulturist-cum-naturalist Ernst van Jaarsveld criss-crossed South Africa, collecting not only carefully documented plant specimens, but also the rocks and soils of the full range of substrates that the country's succulents need for growth. The resulting conservatory is both a botanical and geological museum, and more. Mindful of the fact that South Africa's succulents occur in both western, winter-rainfall and eastern, summer-rainfall zones, the Conservatory's mini-ecosystems are oriented by longitude and latitude according to the geographic distribution of the species displayed. Careful watering by hand ensures that both the amount and seasonality of irrigation is appropriate to the eco-region represented. So successful is the creation of differing zones that visitors in summer might wonder why half the display (from the winter-rainfall region) looks dormant, while the other (from the eastern, summer-rainfall region) is lush and in full flower.

Designed by architect Julian Elliot, the Botanical Society Conservatory has elegant lines and highly efficient passive air conditioning.

126 KIRSTENBOSCH

The Conservatory's centrepiece, a large Baobab Tree, was collected in the Limpopo valley and transported 2 000 kilometres from the far north of South Africa to Kirstenbosch Garden at its southwestern tip.

In addition to the focus on arid-zone succulents, the Conservatory has four specialist collections housed in the four corners of the Conservatory: moist forest ferns, clivias and streptocarpus provide the luxuriant show familiar in greenhouses; a second collection displays a seasonal parade of spectacular bulbs; a third is devoted to the tiny lithops (or stone plants) of the Knersvlakte; and the fourth collection, recently established, showcases the unique *Welwitschia mirabilis* of the Namib Desert of Namibia and Angola. Welwitschia seeds have successfully been germinated and grown at Kirstenbosch for many years. The trick, it has been discovered, is to treat the seedlings, which are extremely vulnerable to fungal infection, with an inoculant that forms a symbiotic association with the roots of welwitschia, preventing attack by other fungi. The favourite plant of the Garden's founder, Harold Pearson, is now being given centre stage.

Maintaining such a diversity of species in semi-artificial conditions requires special skills. To avoid the use of chemical insecticides in the Conservatory, biological controls have been introduced. Gecko shelters were built for the thick-toed gecko *Pachydactylus bibronii*, which feeds on moths, thus reducing caterpillar damage. Predatory mites *Ableseius californicus* were introduced to control red spider mites; the parasitic wasp *Encarsia formosa* to control white fly; and ladybirds to control aphids. Even slug-eating snakes *Duberria lutrix* help by feeding on snails. Sunbirds pollinate the aloes, and Karoo Prinias assist in insect control. Geology, climate, botany and ecology meet in a single space.

A Garden for all seasons

The majestic national flower, the King Protea Protea cynaroides *is always a favourite among visitors.*

Looking beyond fynbos

As the first 'national' garden in South Africa, Kirstenbosch has, from its founding years, attempted to display as many as possible of the country's 20 456 indigenous fern and flowering plant species. We have seen that climatic limitations have not defeated the horticulturists' skills – but space has been a challenge. Today the Garden's living collections include over 6 000 species. The key fynbos families have been given special attention, but so, too, have other families that are more widely distributed in South Africa.

Pearson's passion for cycads received early expression in the collection he established at the Garden's focal point: probably the world's most comprehensive collection of the genus *Encephalartos*, it represents all South African species. Careful hand-pollination ensures a good crop of seeds, regularly distributed to other gardens around the world. Studies on the genetic diversity within the living collection of some rare cycad species growing at Kirstenbosch have shown that a richer gene bank is held in the Kirstenbosch collection than that of the few remaining populations surviving in the wild. As naturally occurring populations go extinct, Kirstenbosch offers an ultimate gene bank to save these 'living fossils'.

Restoring the forests

A necklace of isolated patches of moist, Afrotemperate Forests, starting in Ethiopia, stretches southwards through East Africa, Malawi, Zimbabwe, Mozambique and the KwaZulu-Natal escarpment, ending in the ravines of Kirstenbosch. Here remnant natural forests in the Nursery Ravine and Skeleton Gorge protect some of the southernmost outliers of the rich Afrotemperate Forest flora. The forests on Ruwenzori, Kilimanjaro and Mount Kenya, and many of the mountains of the eastern escarpment, carry a similar but much richer forest flora to these relict patches in Kirstenbosch, which remain as reminders of an earlier, cooler and moister climate. Only 33 tree species occur naturally in the Kirstenbosch forests, but several of them – Cape Beech *Rapanea melanophloeos*, African Holly *Ilex mitis* and White Pear *Apodytes dimidiata* – are common to forests stretching from the equator to the Cape Peninsula.

Although Kirstenbosch entered recorded history as 'Leendertsbos', implying a forested estate, and despite early efforts to protect these forests, little remained by the end of the 18th century of the fine trees that once provided a dense canopy and the special microclimate needed to sustain the forest fauna and flora. Invasive trees, shrubs and climbers filled the gaps left by the woodcutters. Following a 50-year programme to remove invasive alien plants from the forests, initiated by Jack Marais (Curator 1959–1979), recovery has been impressive. Leading the recovery were pioneer species such as Keurboom *Virgilia oroboides*, Tree Fuchsia *Halleria lucida*, Turkey Berry *Canthium inerme*, Wild Pear *Kiggelaria africana* and Cape Myrtle *Myrsine africana*. Good specimens of Ironwood *Olinia ventosa*, Wild Olive *Olea capensis*, Bastard Saffronwood *Cassine peragua*, Assegai *Curtisia dentata*, Bladdernut *Diospyros whyteana* and Rooiels *Cunonia capensis* now clothe the cool, moist ravines above the main Garden. New plantings of the most sought-after cabinet woods, such as Real Yellowwood *Podocarpus latifolius* and Black Stinkwood *Ocotea bullata*, have yet to replace the mighty trees felled in the 18th century.

Forest streams burst through the undergrowth after an autumnal downpour.

A Garden for all seasons

130 KIRSTENBOSCH

ABOVE *Bulbinella latifolia* var. *doleritica is one of the 309 species of geophyte found on the Bokkeveld Plateau, making the Nieuwoudtville region the 'Bulb Capital of the World'.*

FROM FAR LEFT TO RIGHT
Ornithogalum dubium, Romulea monadelpha, Hesperantha viginata and Geissorhiza radians flower in the Bulb House of the Botanical Society Conservatory.

Bulb capital of the world

Most travellers in the northwestern Cape pass through the sleepy little village of Nieuwoudtville without stopping. But the area is now popularly referred to as the 'Bulb Capital of the World': the epicentre of South Africa's unequalled richness in bulb species, with no fewer than 2 200 species occurring in the region. By comparison, California has 250 species, Chile 250, Australia 750, and the whole Mediterranean Basin some 1 335 species.

In 2007, the Hantam National Botanical Garden was established at Nieuwoudtville, the ninth NBG in the national network (see page 230). Before this, horticulturists at Kirstenbosch were faced with the challenge of *ex situ* conservation of the many rare and endangered species of the Cape's stunningly beautiful bulbs. A few genera (*Agapanthus*, *Dietes*, *Eucomis*, *Chasmanthe*, *Clivia*, *Gloriosa* and *Watsonia*) are able to endure the heavy winter rainfall in Kirstenbosch if planted in well-drained soils – and are fortunately not attacked by the Garden's healthy populations of molerats and porcupines. But most bulbs constitute the preferred diet of these Garden inhabitants, and cannot be planted out in the beds. Mathews was to learn the hard way: in a single night in 1916 he had 100 metres of newly planted bulb beds destroyed by a single pair of visiting porcupines.

Kirstenbosch bulb expert Graham Duncan has built a scientific collection of over 900 bulb species, mainly spring flowering. Of these, 221 are threatened in the wild. To protect them from the predations of molerats and porcupines, most are grown in pots in the Garden's large Millennium Glasshouses, where the scientific and *ex situ* conservation collections are held. As they come into flower, the bulbs are displayed in the Conservatory, a 'must-see' in spring and autumn. Because they are grown in bulk in the greenhouse, potted bulbs can also be displayed in the Garden, with the pots being 'plunged' in autumn and lifted after flowering in late spring, and returned to the nursery for the summer dormant period.

The Hantam NBG area is not only the Bulb Capital of the World; it also lies adjacent to the heart of another botanical spectacle – Namaqualand. Flower tourism in South Africa is strongly promoted by scenes of the spring flowers of Namaqualand. Fields of daisies stretch from horizon to distant horizon, with palettes of white, yellow, orange and mauve, and scattered splashes of red and blue. Kirstenbosch creates its own spring displays by the careful selection of species adapted to its rather moist, and less sunny, climate. An explosion of colour is presented in September – felicia, euryops, senecio, cotula, dimorphotheca, osteospermum, ursinia, arctotis, gazania, arctotheca and many more.

While Pearson is rightly honoured for his vision and determination to establish Kirstenbosch, it was Joseph Mathews, more than anyone else, who gave this modern Eden shape and colour. It is fitting to conclude this chapter with a quote from his farewell speech:

'The work at Kirstenbosch should be a labour of love in the greatest job ever begun in the botanical world and I feel that it is an honour to have had anything to do with that job'.

A Garden for all seasons

A bird-watching destination

With 139 recorded bird species, Kirstenbosch is a favourite destination for bird watchers. The commonest breeding residents include the Cape Batis, Cape Robin-Chat and Cape Spurfowl. Ever popular with visitors, but not indigenous to the Western Cape, is the Helmeted Guineafowl. A pair of Spotted Eagle-Owls breed annually in the Garden, their fluffy offspring a source of delight for visitors. Early morning visits will be rewarded by the dawn chorus of the Sombre Greenbul, Olive Thrush and Cape Canary. One of several 'specials' that bird watchers look out for is the Knysna Warbler, an elusive endemic that is becoming increasingly rare due to the reduction of its preferred bramble habitat. Other Kirstenbosch favourites include the Orange-breasted Sunbird and Cape Sugarbird.

CLOCKWISE FROM TOP RIGHT *Spotted Eagle-Owl, Cape Batis, Helmeted Guineafowl, Knysna Warbler*

CLOCKWISE FROM TOP LEFT *Cape Spurfowl, Cape Robin-Chat, Cape Canary, Sombre Greenbul*

A Garden for all seasons

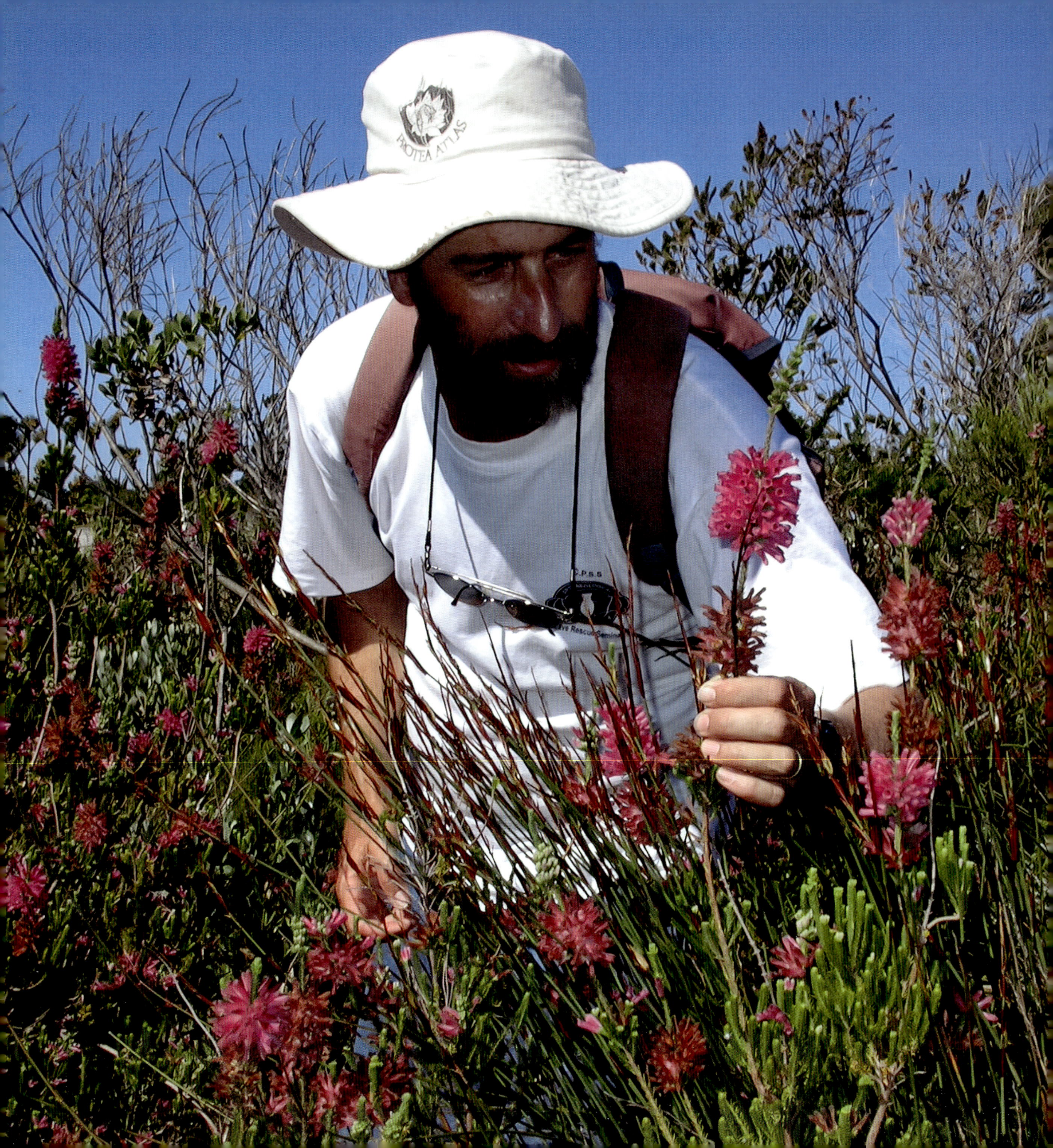
PROTEA ATLAS

CHAPTER SEVEN

Conservation through cultivation

An integrated approach

The vegetation of South Africa is the richest in the world, not only as to the number of species but also as containing an astounding variety of plants of special and peculiar type … special and peculiar plants are sure to be swept out of existence altogether unless special provision is made for their preservation. In future centuries hundreds of these beautiful and remarkable plants will be unknown save by dried specimens preserved in State Herbaria.

BARON FERDINAND VON MUELLER
Famous Australian botanist, in correspondence with Sir Hercules Robinson, British High Commissioner at the Cape, 1895

Kirstenbosch Collections Manager Anthony Hitchcock examines a restored population of Erica verticillata *at Rondevlei Nature Reserve.*

The problem – botanical gardens or museums of the living dead?

The primary objective of botanical gardens is to establish collections of living plants in accord with aesthetic, scientific and educational considerations, presented in displays accessible to the public. Increasingly, emphasis has been given to the inclusion in these collections of 'species of conservation concern', more commonly termed 'threatened species'. For many decades, however, the pursuit of species collections was little more than botanical 'stamp collecting'. Gardens around the world compiled impressive lists of the species in their collections, and distributed their Index Seminum, or seed list, to peers around the globe in a somewhat boastful exercise. The greater the number of species listed, the greater the importance of the garden.

During the 1980s, conservation organisations such as the International Union for the Conservation of Nature (IUCN) developed long-term strategies for saving the world's biodiversity. Vernon Heywood of Reading University, Peter Raven of Missouri Botanical Garden, David Bramwell of the Canary Islands, and colleagues from IUCN, established Botanic Gardens Conservation International (BGCI). Kirstenbosch was a founder member of BGCI, hosting meetings in Cape Town in 1988 and 1998. BGCI, under the leadership of Vernon Heywood and, later, Peter Wyse-Jackson, initiated discussions leading to the approval in 2002 of the Global Strategy for Plant Conservation (GSPC) by the Convention on Biological Diversity.

This short history serves to highlight the processes that shaped the policies and corporate plans of Kirstenbosch, the NBI and, later, of SANBI. Following the political transition of 1994, South African conservation scientists were once again able to participate in and benefit from the rapid changes in thinking and action resulting from the Earth Summit that had been convened in Rio de Janeiro in 1992. One of these changes was the move away from the 'preservation' to the 'conservation' approach to managing living collections of plants.

In the language of botanical gardens, a distinction is made between *in situ* and *ex situ* conservation. The first

The Millennium Glasshouses, sponsored by the Botanical Society, house the extensive scientific collections of Kirstenbosch – the ex situ *living collections that are of critical importance to plant conservation.*

relates to achieving conservation goals by protecting threatened species and associated plants and animals in their natural habitat. *Ex situ* refers to keeping specimens of such species in the living collections of botanical gardens. These living collections are most often kept 'off limits' to visitors – to ensure that they can be carefully managed and are not vulnerable to theft, as many threatened species are of great rarity and high value to commercial collectors.

In reality, *ex situ* approaches have serious limitations and the extensive living collections in Kirstenbosch and the other SANBI gardens have been reviewed and restructured because of this. Firstly, with a flora of over 20 000 species, of which 25 per cent are of 'conservation concern', there is simply not enough nursery space to house even a small proportion of such species. Secondly, living collections need highly specialised horticultural skills to manage. Thirdly, many of the most threatened species are difficult to grow, short-lived or disease-prone under nursery conditions. And, finally, the very small sample populations that can be maintained in living collections are not representative of the great genetic variability of wild populations.

Globally, botanical gardens are now co-operating in programmes aimed at sharing responsibility for securing the survival of threatened plants: each garden focuses on a few groups of species that can best be maintained in living collections until their re-introduction to the wild can be implemented. Several approaches are being tested, with Kirstenbosch taking a leading role.

The solution – the Threatened Species Programme

When Anthony Hitchcock was appointed Collections Manager at Kirstenbosch in 2000, he was faced with a serious dilemma. The Garden had an enviably large collection of rare species of many genera, all individually potted and neatly arranged in various greenhouses: *Ericas*, *Pelargoniums*, *Disas*, *Streptocarpus* and many more. But there were very few horticulturists with time available to care for the massive collection. It was impossible to cover even a fraction of the over 1 000 species known at the time to be under threat in the Cape Floral Kingdom, and more were still to be added to the Red List as detailed assessments progressed. Hitchcock's challenges included: how to conserve so many threatened plant species effectively, with a good representation of their genetic diversity? For how long can these plants be kept in effective *ex situ* collections, and how can they be accommodated? How can the conservation message be communicated to policy- and decisionmakers, and to the general public?

The Global Strategy for Plant Conservation provided Hitchcock with some guidance. The GSPC was unique among similar conservation agendas in that it set measurable targets for its objectives. Target 8, for instance, states that 'by 2020, there should be at least 75 per cent of threatened plants in accessible *ex situ* collections, preferably in the country of origin, and 20 per cent of them available for recovery and restoration programmes'.

A decision was taken to focus on threatened habitats, and to implement restoration projects that had a good chance of success, so ensuring the long-term protection of the target species. As in so many matters, Pearson, in his 1911 paper, was ahead of his time when he observed: 'Protection, to be effective, must be applied, and strictly applied, to areas, not to selected species.'

The Garden of Extinction displays species that have been extirpated in the wild, or are Critically Endangered.

Conservation through cultivation

Plant labels provide information on the species displayed in Kirstenbosch – such as this locally extinct Spiderhead Protea.

A few 'flagship' habitats and species were chosen as case studies that could be replicated as experience developed.

A second leg of the Threatened Species Programme (TSP) was to overcome the challenge of limited resources. Instead of being a strictly controlled 'Kirstenbosch' project, the TSP was opened up to multiple partners – with a rapid expansion of resources, skills and ownership. New role players included CapeNature and South African National Parks, the City of Cape Town, Working for Wetlands, the Biodiversity and Wine Initiative, the Botanical Society, civic organisations, private landowners and many more.

A third leg was to practise *ex situ* conservation at Kirstenbosch in combination with seed banking. Kirstenbosch had, since its early years, offered surplus plants and seeds to members of the Botanical Society, and for many years the Society's annual seed list was eagerly awaited by its members. By the 1980s it was evident that merely distributing seeds to home gardeners was of little benefit to conservation, and that the Seed Room at Kirstenbosch did not have the facilities to act as a seed bank for the long-term preservation of seeds. Kobus Eloff, during his term as Director, attempted to establish a proper seed bank but, even though a brave start was made, the required funding was not available.

It was therefore fortuitous that, when Kew established its multimillion pound 'Millennium Seed Bank' in 1998, one of the first partners they approached was Kirstenbosch. Sir Ghillean Prance, then Director of Kew, was a long-standing Kirstenbosch friend, and an agreement was soon reached with NBI to collaborate in collecting viable samplings of 10 per cent of all South African species for deposit in the huge seed bank established at Kew's sister garden at Wakehurst Place, West Sussex. Several Kirstenbosch horticulturists received advanced training at the Millennium Seed Bank (MSB), and Kirstenbosch led the international network of teams in reaching the goal of 10 per cent of the national flora in record time. As a globally shared resource, the MSB offers South Africa first call on the valuable collection of our seeds safely stored in a world-class facility at no cost. This is a win-win situation for plant conservation.

Increasingly, the Kirstenbosch business model, and that of NBI and SANBI, has become one of 'strategic partnerships'. The perennial problem of limited resources can be resolved only through collaboration, and nowhere has this been more evident than in dealing with plants that are facing extinction – or even plants believed to be extinct.

Within a few years, the TSP had become a major programme, and proved to be a remarkable success. And, as is well known, success breeds success.

The Kirstenbosch Threatened Species Programme has 23 Cape lowlands species in restoration programmes – two of which are extinct in the wild. One of the most remarkable and exciting stories to come out of Kirstenbosch in recent years was that of the beautiful Whorled Erica – *Erica verticillata*.

Encephalartos hirsutus, probably the most seriously threatened plant in South Africa, is now reduced to a single population.

The fynbos-endemic Orange-breasted Sunbird pollinates the flowers of a restored population of Erica verticillata.

Extinction, rediscovery and reintroduction

In chapter 4 we discussed the European plant collectors Franz Boos and Georg Scholl who, in the 1780s, collected impressive numbers of plants that were soon flowering in the gardens of Schönbrunn and Belvedere palaces in Vienna. Remarkably, many of these species survived the turmoil of wars and political upheaval for over two centuries, while back home, some have not fared so well – and, perhaps least well of these, *Erica verticillata*, a hardy, tall-growing erica with splendid mauve-pink flowers.

The sandy plains of the Cape Flats were the natural habitat of this erica. A wilderness in the 1780s, the area is now almost totally transformed by urban and agricultural development. Indeed, today the Cape lowlands constitute one of the most threatened ecosystems of South Africa. Detailed surveys of land transformation for the entire country were conducted by Mathieu Rouget and colleagues at the Kirstenbosch Research Centre during 2004. These concluded that only 19 per cent of Acid Sand Plain Fynbos remains, and none of this is formally protected. In contrast, the Cape Peninsula mountain fynbos has 94 per cent intact and 90 per cent protected. By 2011 the area of undisturbed Acid Sand Plain Fynbos had dropped even further, to less than 14 per cent of its former range.

E. verticillata was last known from a herbarium specimen collected on the Cape Flats in 1908 – never to be seen in natural populations again. Deon Kotze, while erica specialist at Kirstenbosch in the 1980s, tracked down surviving specimens in a municipal garden in Pretoria, and at Kew. Yet another specimen – possibly a remnant from plantings during the earliest days of the Garden – was found in 1990 in an overgrown area in the middle of Kirstenbosch by Garden foreman Adonis Adonis.

So, by 1990, three clones of *E. verticillata* were available for propagation and possible re-introduction to their natural habitat. When Anthony Hitchcock followed up on this enigmatic species, his colleague Ted Oliver, erica specialist at the Compton Herbarium, recalled having seen *E. verticillata* at the gardens of Belvedere Palace in Vienna during a visit in 1967. Hitchcock visited these collections in 2005, and found no fewer than 20 species of erica growing happily in the nursery – many of these now rare and threatened in their natural habitats, but having survived at Belvedere since the 1780s.

With some high-level support from the South African Ambassador to Austria and the Austrian Minister of Agriculture, the many bureaucratic hurdles to repatriating some living material of *E. verticillata* were overcome. When the material eventually arrived in Kirstenbosch, it was found to represent two clones, one pink and one red. Molecular studies at the Kirstenbosch Research Centre and the University

Members of the Millennium Seed Bank team return from a productive collecting trip in Mountain Fynbos.

140 KIRSTENBOSCH

of Cape Town have confirmed that the living collection includes five distinct clones of E. verticillata. Today Kirstenbosch has a robust gene bank of this species, for nearly a century recognised as 'extinct in the wild'.

Having built up a collection of E. verticillata, the next step was to reintroduce the plants to their natural habitat. Pioneering work on reintroduction undertaken by Dalton Gibbs of the Rondevlei Nature Reserve was followed up by a team drawn from the City of Cape Town, CapeNature, the University of Cape Town, Kirstenbosch and the owners of Kenilworth Racecourse – surprisingly, Cape Town's premier racecourse provided one of the few intact areas of Acid Sand Plain Fynbos that remain in a highly urbanised area.

After a promising site for reintroduction had been found, with a signed Memorandum of Understanding between all partners, more hurdles had to be overcome. The 42 hectares of land in the centre of the racecourse had not been burnt for over 100 years. Despite this, the site is home to 271 species of indigenous plant, of which at least 19 are included in the Red List. The ecologists and horticulturists all agreed that, prior to reintroduction, the land should be carefully cleared of all alien species and then burnt. But on the agreed date for the burn, in February 2004, an outbreak of horse influenza occurred and the horses stabled on the racecourse could not be moved from the quarantine stables, situated near to the path of the planned fire. And so the burn had to be postponed until March 2005.

A specimen of Erica verticillata *was found in Kirstenbosch by Garden foreman Adonis Adonis in 1990.*

The seeds of threatened species such as Erica verticillata *are propagated at the Kirstenbosch seed orchard for restoration purposes.*

The project was not derailed by all these challenges, and today a healthy population of E. verticillata is once more flowering in sheets of mauves and pinks, pollinated by Orange-breasted Sunbirds (endemic to the fynbos), in the middle of one of the country's busiest racecourses.

The living collections at Belvedere Palace, Vienna, show that species can survive apparently indefinitely away from their natural habitats, in ex situ collections, and be successfully re-established. As a bonus, the propagation programme at Kirstenbosch has been so successful that the species is now available to the public for purchase – in several shades of pink – at the Garden's annual Plant Fair.

Conservation through cultivation

Spring annuals have always been a special feature of Kirstenbosch. Yellow and orange Ursinia anthemoides *and purple* Senecio elegans *crowd the foreground. Other daisies include the light yellow* Arctotis acaulis, *cream* Arctotis hirsuta *and* Dimorphotheca pluvialis.

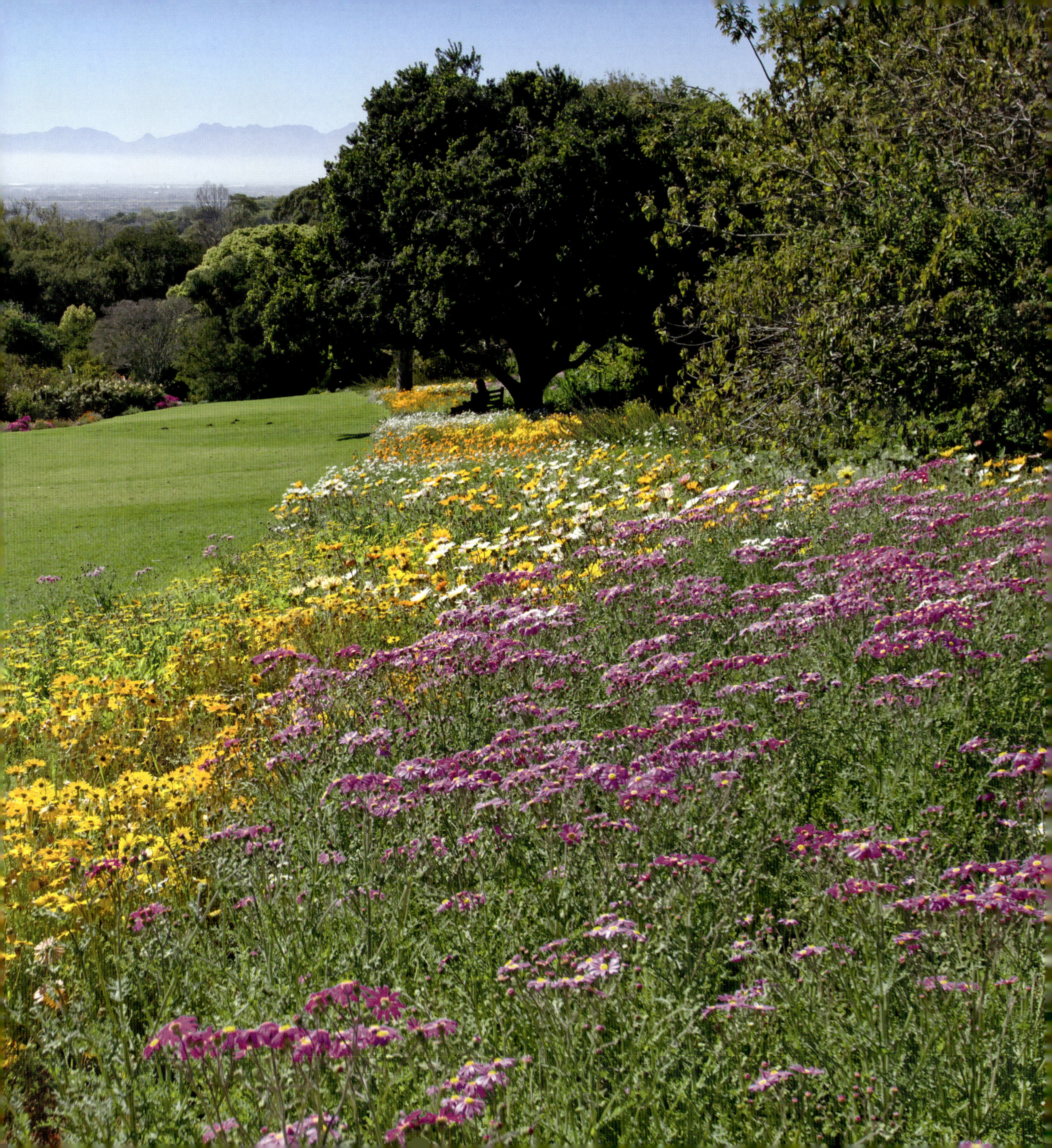

Taking flowers to the people

We have already noted the passion for South African plants displayed by the royal gardeners of Europe at the close of the 18th century. No surprise, then, that Pearson, writing in 1910, was greatly disturbed by the apparent lack of interest shown by South Africans in their flora:

> '... even to-day, you will find at Kew, at Dahlem and I believe also at Edinburgh, Vienna and elsewhere in Europe, a greater variety of South African plants than in the gardens of the Cape itself'.

Perhaps this situation influenced his decision to restrict the plantings at Kirstenbosch to indigenous species, and so popularise the use of the flora in home gardens, as they were already being so widely used in Europe. Pearson's criticism was not entirely fair, as some enthusiasts had already made efforts to display the Cape flora to local audiences. By the time of his arrival in the Cape, several villages had initiated annual wild-flower shows. First of these was Tulbagh, in September 1887, a country village now surrounded by extensive wheat fields and vineyards but then home to an extraordinary diversity of spring-flowering bulbs – gladiolus, ixia, watsonia, sparaxis, freesia, geissorhiza, ornithogalum, plus many ericas, proteas, daisies, etc. One newspaper correspondent described the scene:

> 'The fields and mountains are gaudy to profusion in their spring habiliment, and flowers numerous as the stars in the firmament are spread about up hill and down dale in the grass and bushes.'

Encouraged by the success of the first show, the following year the village organising committee attracted 'the distinguished patronage vouchsafed by Lady Robinson, who has graciously signified her intention of being present'. Lady Robinson was the wife of Sir Hercules Robinson, Queen Victoria's High Commissioner to South Africa, and a colonial official of considerable stature. A special train was arranged for the convenience of the party from Government House – a measure of the personal interest in matters botanical that

The annual Caledon Wild Flower Show celebrated its centenary in 1992 – the longest unbroken record for wild-flower shows in the world.

Apart from winning 33 gold medals at Chelsea since 1974, the Kirstenbosch exhibit at Hampton Court Palace won gold in 1995.

Elandsberg Nature Reserve

Despite the transformation of over 80 per cent of the fynbos, and especially the renosterveld of the Tulbagh area, to wheat and other agricultural uses, new species continue to be found in the tiny remnant patches of natural veld. A single privately owned property, Elandsberg Nature Reserve, has yielded several new species over the past decade, most of these named by John Manning of the Compton Herbarium.

Spring flowers erupt following a recent veld fire in the renosterveld of Elandsberg Nature Reserve.

was entertained by many of the political leaders, and those in business, of the Victorian era. Governors General, Lord Chief Justices, Randlords and even Prime Ministers – in the case of Jan Smuts – were personally involved in field expeditions, herbarium collections and the creation of new botanical gardens. Sadly, little of this rich tradition remains.

While Tulbagh holds the honour of being the first village to organise a wild-flower show, it was Caledon that became the focal point of such enterprises, starting in 1892. Like Tulbagh, Caledon lies in the centre of rich wheat lands on the Bokkeveld shales that provide the fertile soils on which renosterveld, as opposed to fynbos, is dominant. The adjoining mountains, of acidic, nutrient-poor sandstones, carry fynbos. These two major vegetation types have distinctive and extremely rich endemic floras and, in spring, local farmers could readily supply up to 400 species of wild flower, in profusion, for the massed arrangements that became the hallmark of the wild-flower shows of the Cape.

The transformation of renosterveld into wheat lands, and the rigid protection of mountain fynbos within nature reserves, have reduced the source of abundant flowers for shows today. But the rise in the commercial cut-flower industry in the Cape provides an adequate supply of flowers for annual spring shows that are organised by local communities from Clanwilliam in the north, through the villages of Hopefield, Darling, Caledon and Hermanus, to Bredasdorp and Riversdale in the far south. Fortunately, the timing of shows follows the southward progression of the flowering peak, from August in the north to October in the south, allowing flower lovers to move from one show to the next through the spectacular Cape spring.

Over the years, major flower shows with participation from across the country have been held in Cape Town – at first in the City Hall and, when the demand for space outgrew the venue, at show grounds and conference centres around the city. Kirstenbosch has organised a series of flower extravaganzas to celebrate special anniversaries, such as its Golden Jubilee of 1963, and 75th Anniversary in 1988, when tens of thousands of visitors enjoyed a botanical spectacle that was world-beating in terms of colour, fragrance and diversity. Grand floral shows probably reached their zenith in the great *Flora '93*, when 65 000 excited visitors enjoyed a mass display of the country's floral riches. As the pressures on natural populations of wild flowers increase, and the energies of a remarkable cohort of volunteer organisations wane, the possibility of mounting shows such as those of the past, more gracious, century has disappeared, although smaller shows in country villages continue.

Spring annuals from Namaqualand greet visitors on arrival in the Garden.

Chelsea Flower Show

The early collectors introduced to the world the botanical marvels of the Cape, but Kirstenbosch was not placed firmly on the horticultural map until Director Brian Rycroft promoted the Garden at international flower shows in the United Kingdom, the Netherlands, Monaco and the United States. Nowhere has the contribution of exhibits from Kirstenbosch been more successful than at the Royal Horticultural Society's Chelsea Flower Show.

Chelsea is rightly regarded as the world's premier event of gardening excellence. The Kirstenbosch exhibit was, for many years, financed by the Ministry of Foreign Affairs, and built by an English flower arranger. From 1990, the Ministry withdrew its funding, and a private sector sponsor had to be found. A plea for support placed in the *Cape Times* drew an immediate response from Ian Sims, Chairman of Syfrets, an old Cape Town financial institution, and the long tradition of a Kirstenbosch exhibit at Chelsea could be continued. To make the project truly South African, David Davidson, graphic designer at Kirstenbosch, was chosen to lead the team. With the assistance of Raymond Hudson, a Johannesburg horticulturist, Davidson created a winning exhibit, won a gold medal, and has continued an unbroken record of success ever since. In the past 37 years, the Kirstenbosch exhibit has won 33 gold medals, several silver gilt medals and one silver medal. More importantly, Chelsea has attracted thousands of garden tourists to the Cape, and countless friends and supporters to the Garden.

Kirstenbosch has represented South Africa at the Royal Horticultural Society's Chelsea Flower Show for 37 years, bringing home 33 gold medals for exceptional exhibits – such as these displays, seen at recent shows.

Conservation through cultivation 147

Plant genetic gold

The most successful, but poorly acknowledged, export from South Africa, besides gold and diamonds, is the genetic material on which the massive 'geranium' pot-plant industry of Europe and other regions is based. Early collectors at the Cape amassed vast collections of plants for transfer back to the gardens of Europe. None proved more responsive to cultivation and hybridisation than members of the genus *Pelargonium*. Of the 220 species in South Africa, 119 of which are in the Garden, about 80 per cent are confined to the winter-rainfall area of the Cape. Hybrids of the original gene pool adorn nearly every garden and flower box in Europe. That not a cent in royalties has ever been paid for this national genetic resource is a matter of irritation to many who believe in the hopes and principles of the Convention on Biological Diversity. But the 'low-hanging fruit' of South Africa's rich horticultural resources had been plucked long before modern concepts of intellectual property and sovereign ownership of genetic resources had been placed on the political agenda.

Many thousands of pelargonium hybrids and cultivars – gardening favourites for over 200 years – have been bred by British horticulturists.

Plant exports – the good, the bad and the ugly

While we are increasingly aware of the effects of alien invasive species in South Africa, especially Port Jackson wattles, Rooikrans, hakeas and pines that infest both lowland and mountain fynbos, we are often unaware of what we, too, have given the world in aggressive invaders.

We have already noted the interest of royal gardens in our flora – an 18th century passion for plants that became a driving fashion in middle-class European homes in the 19th century. The pelargoniums sent to Kew by Francis Masson were soon being hybridised, creating a stunning variety of colour and flower forms, and being planted in millions of window boxes in homes and village squares across Europe. The popularity of our mesems, nemesias, streptocarpus, gladiolus, watsonias, freesias, ixias, dieramas, agapanthus, arctotis, gerberas and strelitzias, and so many more, has led to some of these plants being considered native in their new countries – even adopted as city icons, as in the case of *Strelitzia reginae,* the floral emblem of Los Angeles.

But not all of our plants have remained welcome in their new homes. Who would have thought that the beautiful arum lily, prized by so many for festive occasions in Europe, would become a plague in New Zealand? Or that one of our most threatened species *Gladiolus caryophyllaceus* would raise the ire of residents of Perth, Western Australia? These and others such as the ice-plants, agapanthus, bietou and veld grasses are viewed by some as invasive aliens.

During the 1990s, Kirstenbosch hosted many meetings of an international team of experts that initiated the Global Invasive Species Programme (GISP), which is aimed at preventing – or at least reducing – the spread and impact of aggressive alien plants and animals around the world.

The passion with which some people react to alien species suggests they have forgotten the good intentions behind their introduction. In 1880, the Colonial Forester planted pines on the slopes of Table Mountain to 'cover the mountain's naked appearance ... an offence and eyesore to Cape Town'. It was Joseph Mathews himself who introduced the now infamous kudzu vine to Kirstenbosch, remarking in his Annual Report for 1919: 'A few plants of the Kudzu vine have been planted out', and in 1920 'This season's growth of the Kudzu vine has reached 30 feet'. Fortunately, the vine died soon after this report.

One of the leading personalities in the GISP team was Kathy MacKinnon, ecologist at the World Bank, and another of Kirstenbosch's many influential friends. MacKinnon was quick to remind us of our contribution to the global problem of invasive species, and the Kirstenbosch team responded promptly with a fascinating, and telling, theme garden on 'Weeds South Africa gave the world' – a sobering reminder of the dangers inherent in plant introductions from distant lands, as Pearson had so perceptively pointed out in 1910:

> *'It is hardly necessary to say that there is a risk in introducing a new plant from another region. It might prove itself to be a good servant, but there is always the possibility that, if it is not carefully looked after, it may become a bad master.'*

Two of the worst invasive plants originating in South Africa are now serious weeds in their adoptive lands: in California, Sour Fig Carpobrotus edulis (ABOVE), *and in Australia, Bietou* Chrysanthemoides monilifera (RIGHT).

Conservation through cultivation

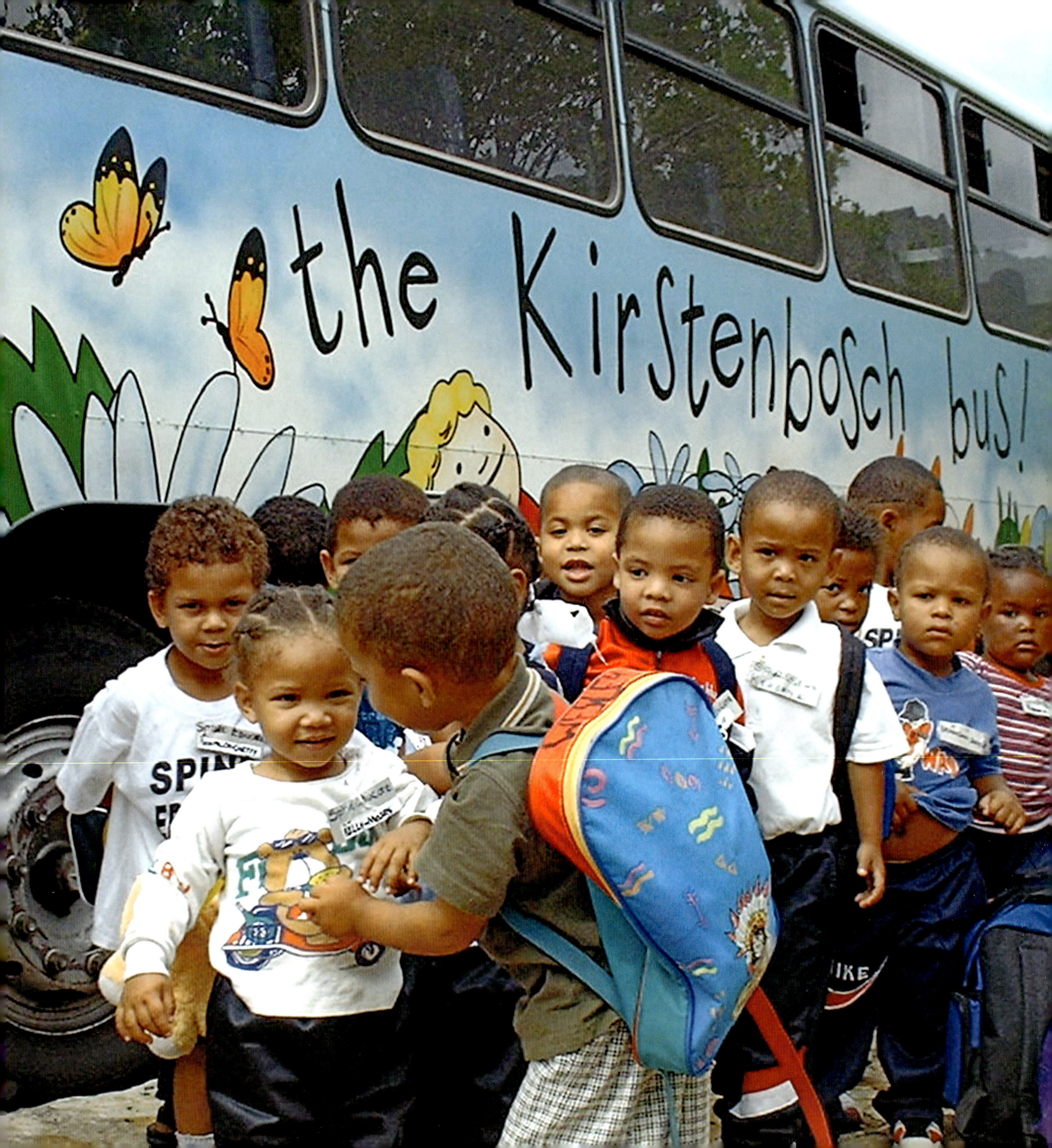
the Kirstenbosch bus!

CHAPTER EIGHT

Come smell the flowers

Inspiration and education

Environmental education has been described as a response to the environmental crisis. But in South Africa, with its long history of unequal provision of educational resources, we could also view environmental education as a response to the education crisis.

ALLY ASHWELL
Veld & Flora, *1997*

The Kirstenbosch bus has brought over 200 000 school children to Kirstenbosch since it was first sponsored by the Anglo American and De Beers Chairman's Fund in 1996.

Two remarkable women

Kirstenbosch has always been at the forefront of environmental education in South Africa, a proud tradition that has its roots in the first decade of the Garden. The vision so eloquently promoted by Pearson in his seminal papers emphasised the need for a direct link between the Garden and university research and scholarship. But he made no mention of using the Garden as a living schoolroom for the younger generation.

We must pay tribute to Louisa Bolus (née Harriet Margaret Louisa Kensit, 1877–1970) for instigating the use of Kirstenbosch as an outdoor classroom for the school children of the Cape. Miss Kensit started working as an assistant in Harry Bolus' herbarium in 1903, and in 1912, the year after Bolus' death, she married his son Frank, who also happened to be her own father's cousin. This remarkable family made extraordinary contributions to research on the country's flora. As an indefatigable taxonomist who radiated an infectious enthusiasm, Louisa Bolus named no fewer that 1 700 new species, mostly succulents of the family Mesembryanthemaceae – better known as 'vygies' or 'mesems'.

ABOVE *Louisa Bolus, founder of the Kirstenbosch environmental education tradition (Courtesy of the Bolus Herbarium)*

LEFT *Muriel E. Johns (1901–1977) spends time with young well-wishers on the day of her formal retirement in 1959. Her role in promoting Kirstenbosch as an outdoor classroom and inspiring a love for nature is legendary.*

Her role in environmental education is legendary. Soon after the founding of Kirstenbosch, she started taking regular classes of school children on walks around the Garden, teaching them about plants, birds and insects, and their interactions. This was all done on a voluntary basis – she initiated the long and wonderful tradition of volunteerism that has been a mainstay of environmental education in the Garden over its entire history. Compton notes 'it was Mrs Bolus' inspired and inspiring leadership and persistent advocacy which brought about the establishment of a recognised connexion between official school education and Nature Study and all its implications … At Easter 1920 she held a 10-day course for Cape Town Training College students, with far-reaching results.'

These results included the appointment by the Cape Provincial Education Department in 1923 of Laetitia Starke as the first official 'Teacher of Nature Study' at Kirstenbosch, an arrangement that continued unbroken until 1993 – 70 years and several hundred thousand happy school visits later.

Another of Louisa Bolus' outstanding contributions to spreading knowledge of, and promoting interest in, the South African flora is her *First Book of South African Flowers* – a precursor of the Botanical Society's brilliant series of *South African Wild Flower Guides*.

Rustenburg High School visits Kirstenbosch in 1964.

Kirstenbosch was fortunate to have had another remarkable woman appointed as Nature Study Teacher in 1932, the legendary Miss Muriel E. Johns (1901–1977). She continued in this role, even after retirement, until 1962 – a total of 31 years. Building on the programme successfully initiated by Starke, and with the same minimal facilities, Johns presented outdoor classroom experiences for nursery schools, primary schools, high schools and Training Colleges – in fair weather and foul. Anne Bean, a colleague who also served Kirstenbosch for many years, wrote of Johns:

> '… many are the memories, and few are the students that did not retain vivid impressions of her lessons, for she was both a forceful personality and a vivid and dramatic teacher with a flair for making use of educational opportunities unrecognised by others'.

A Lecture Room was built in 1938 (today's Sanlam Hall), providing shelter from the winter storms, and space for a small museum of curiosities – including live frogs, lizards and snakes – to excite and amaze the children. The memories and impressions of the Garden that school children take with them help foster enthusiasm for nature in later life.

Gold Fields Environmental Education Centre

By the 1990s, the changing sociopolitical landscape in South Africa called for new responses to the exciting opportunities offered. In 1993 the then Cape Provincial Administration, using the rather tired excuse of limited funding, terminated the 70-year tradition of supporting two Education Officers at Kirstenbosch. While lack of funds had constrained the Garden's development throughout its history, it now precipitated a new approach.

In 1990 the Corporate Strategic Plan for the newly established National Botanical Institute outlined the mission of the environmental education programme: 'to use the gardens and resources of the NBI to inspire and enable people to take responsibility for their environment'. While retaining the central role of the NBGs as incomparable 'outdoor classrooms', the new concept of 'outreach greening' was accepted – an opportunity for the NBI to contribute actively to the environmental empowerment of the poorer, previously marginalised communities of the townships.

Come smell the flowers

The new vision required both conceptual and physical foundations. Across South Africa, workshops were held to assess the opinions of community leaders in townships – many of these still deeply scarred by the factional violence of the 1970s and 1980s. A strong partnership had already been established with a new NGO, Trees for Africa (TFA), which targeted the environmental needs of impoverished townships. Gradually a coherent programme was crafted, with the technical expertise available in the NBI's gardens, and the fund-raising and social-engagement skills of TFA forming a successful model for moving from garden to community.

An early project involved one of the leading schools of the Cape Flats, Alexander Sinton High School, where Kirstenbosch assisted in establishing an indigenous garden, opened by president-to-be Nelson Mandela in 1992.

The new 'environmental education' model needed an appropriate home base. The Nature Study School, built in 1970, was a rather grim, concrete blockhouse, needing a serious face-lift to make it more welcoming and user friendly. Government funds were not available, but the long association between Kirstenbosch, Cecil Rhodes and Gold Fields of South Africa Ltd, of which Rhodes was a founder, helped secure generous funding from the Gold Fields Foundation in 1992, with R450 000 being made available to renovate the building. An additional grant of R250 000 was received from the Douglas Murray Trust to develop a children's 'Discovery Centre' in the building and to employ two education officers to replace those previously funded by the Cape Provincial Administration. By the end of 1993, a vibrant Environmental Education Centre was operational, with Alice (Ally) Ashwell installed as Education Officer.

School children take a botany lesson in the outdoor classroom.

The stellar Stella Petersen

One of the Garden's many stalwarts, Stella Petersen was the first 'previously disadvantaged' woman to earn a Master's degree in botany from UCT, and was a protégée of Edith Stephens (see also pages 36 and 160). Petersen, whose warmth and charm made her a favourite among the younger visitors, was an active volunteer at Kirstenbosch from the inception of the volunteer programme until her 87th year. Her lifelong contribution as teacher and conservationist was rewarded with an honorary doctorate from her alma mater in 2011. She especially enjoyed the quip of one youngster – 'Granny, you look old, you walk old, but you don't talk old'.

Stella Petersen, veteran volunteer, reminisces with Xola Mkefe, first Black Kirstenbosch 'Outreach Officer'.

An early photograph shows the Nature Study School, built in 1938, and now serving as an events venue – the Sanlam Hall.

One of Nelson Mandela's early associations with Kirstenbosch was the launch of a school greening project at Alexander Sinton High in 1992.

Kirstenbosch was again fortunate to have inspired and passionate leadership, this time to develop its new environmental education programme. Ashwell set about formulating programmes to comply with the new curricula, while ensuring fun-filled, practical, hands-on sessions where learners could engage with indigenous plants and their ecology by touching, smelling, observing, recording and analysing information about them.

Key to the new approach was the development of resource materials and effective preparation of school teachers in advance of visits to the Garden. With only two full-time staff, the programme was mobilised by developing a team of volunteer 'education officers', mostly drawn from people with prior teaching experience and a passion for Kirstenbosch, and who receive intensive training. The first volunteers' course in 1993 had 'no curriculum, no materials, no course outlines, but huge amounts of enthusiasm and a desire to learn'.

Since 1993, the team of volunteers, now called 'Assistant Education Officers' and usually about nine in number, has ensured that Kirstenbosch can offer guided visits to an average of 14 000 school learners per year, plus 6 000 self-guided visitors. The numbers have now reached the optimum level for available facilities, and emphasis can be given to increasing the value for every school child encountering the splendour of Kirstenbosch for the first time.

The spirit of volunteerism is strong, with the Botanical Society having a team of 160 volunteers participating in a wide range of garden activities; they include environmental educators, garden guides, herbarium assistants, library assistants, plant sales organisers and many more.

Students visit the 'Garden of Extinction' to learn about species targeted for restoration projects.

Come smell the flowers

Springtime is alive, with school children flocking to the Garden to 'smell the flowers'. Kirstenbosch is an outdoor classroom of particular splendour and is put to good use by schools throughout the year. Here, the spring flowers make for a colourful backdrop to instruction.

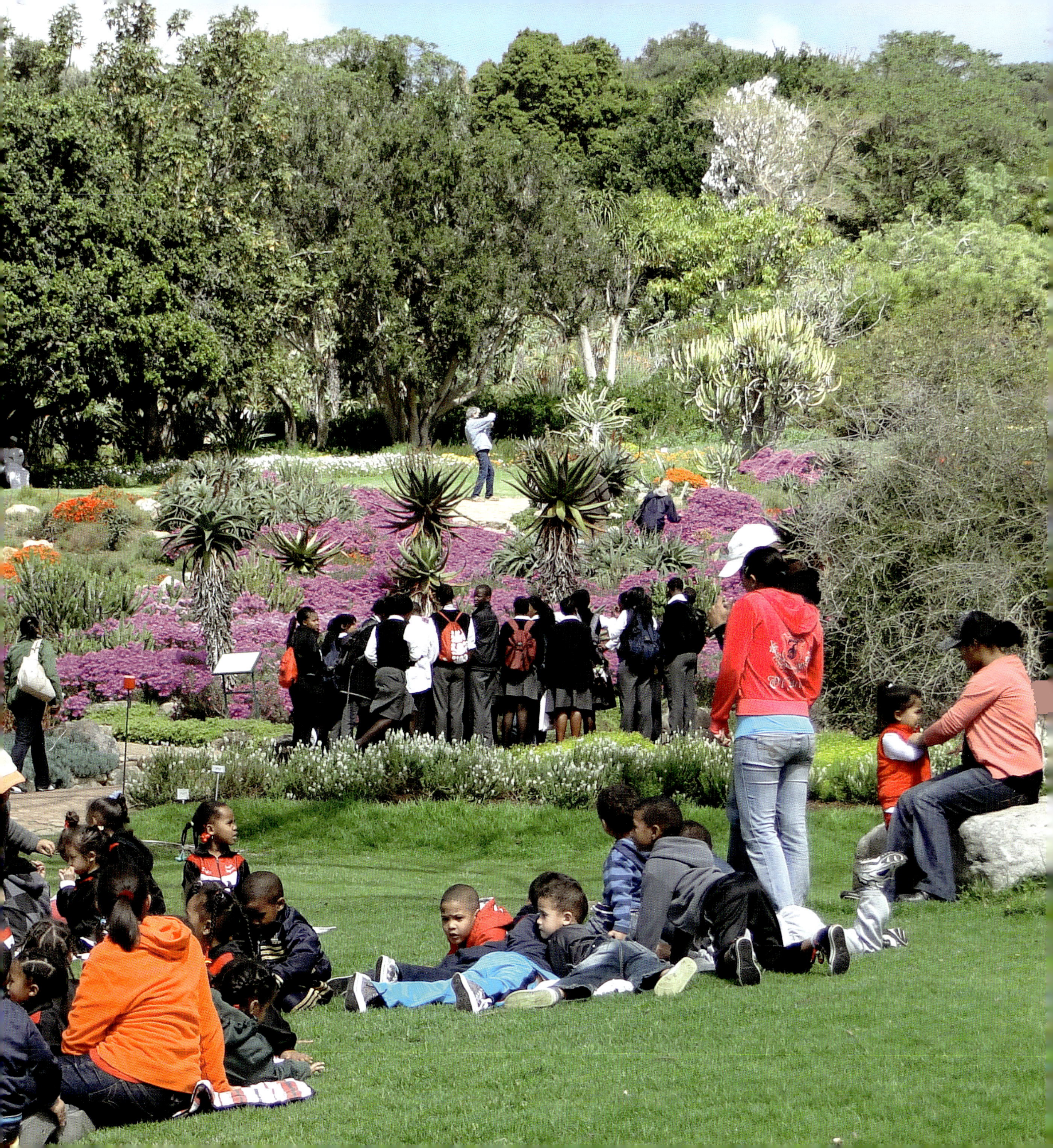

Bringing the community to Kirstenbosch

The early 1990s witnessed an unprecedented growth in fundraising for Kirstenbosch. Contrary to the previous tradition of expecting government to finance every activity, the new leadership in the NBI, strongly supported by the Board, devoted much attention to the development of relationships with benefactors, new and old.

Like Gold Fields of South Africa Ltd, the Anglo American Corporation and De Beers Consolidated Diamond Mines Ltd had their roots in the industrial and business legacy of Cecil John Rhodes. When they were approached in 1995 for a substantial grant of R3.5 million for Kirstenbosch, the response was positive and immediate.

A key requirement was a bus to bring children from the schools of the Cape Flats townships to the Garden. From January 1996, the Kirstenbosch Bus, colourfully painted with flowers and happy children, has become a mobile ambassador for the Garden. Crossing the length and breadth of the Cape Flats – from Athlone to Gugulethu, Langa to Khayelitsha and beyond, on a daily basis – it brings school children, teachers and community groups to enjoy the wonders of Kirstenbosch. Today, some 300 000 kilometres of round trips between the Garden and Cape Flats later, the bus has made – and continues to make – the Garden accessible to communities once politically, economically and spiritually isolated from it.

The Gold Fields Environmental Education Centre serves more than 20 000 learners each year.

Taking Kirstenbosch to the community

In initiating an outreach greening programme, it soon became obvious that the link between the Garden and communities needed to be channelled through the schools. The Kirstenbosch Nature Study School had worked this way, but the link was with the more privileged, predominantly white school community. A far larger, more urgently needy community existed in the townships. This was the new target, to be implemented without neglecting the schools already within the network.

In 1996 the first Black 'Outreach Officers' were appointed. Excellent candidates were selected. An early appointment was Xola Mkefe, whose experience as a teacher and whose skills as an inspirational speaker and a respected figure within the community provided early impetus to the programme. Mkefe saw his role as 'bringing learners to Kirstenbosch, and taking Kirstenbosch to the community'. The activity's popularity among schools led to the rapid growth of the programme, necessitating a constant search for additional funding. Special skills were needed to ensure the project's success in the sandy, nutrient-poor soils of the Cape Flats, and in 1996 the first 'Outreach Greening Horticulturists' were appointed. By 1996, 11 school gardens had been established. In 1999, the Anglo American and De Beers Chairman's Fund provided funds for the appointment of interns as part of the ongoing human capacity development objectives of the programme. In 2001, the Table Mountain Fund took over financing of the bus for several years, with fuel being provided by BP. At each step, the environmental education programme has found willing supporters from the corporate sector, the Botanical Society, WWF, the Table Mountain Fund and many individuals who have donated either time or funds.

Outreach Greening has a strong training element. It is, more than anything, an empowerment process. On-site consultations are followed by a series of workshops and training sessions, supported by handbooks and manuals published by Kirstenbosch and other SANBI gardens and partners. School teachers are trained first in how to plan and design a garden, in soil preparation, plant propagation and basic garden maintenance. Teams of learners are then guided in planting indigenous, water-wise plants that occur naturally in the area. The process is combined with teaching, which follows the formal school curriculum.

By the turn of the millennium, Outreach Greening had become a core programme of the NBI. Major investments in poverty-relief projects within the government's Expanded Public Works Programme provided unprecedented State funding for the type of work that Kirstenbosch had been doing for the previous decade. The 'Greening the Nation' programme was a R70-million project, funded via the Department of Environmental Affairs and Tourism and, from 2004 to 2009, by the SANBI's Environmental Education directorate. It grew out of experience gained in the community-based projects on the Cape Flats, initiated in the first hesitant steps of the early 1990s. Between 1997 and 2011, 69 schools had been incorporated into and supported by Kirstenbosch's Outreach Greening programme, with 27 more funded through the Greening the Nation project.

Growing greener people

'My friends from Nyanga said when they were growing up they thought the mountain was a wall that separated the rich from the poor' – a telling comment made by Zwai Peter, who grew up on the Flats himself and was one of the early champions of a brave idea that evolved into the SANBI Cape Flats Nature Partnership. The partnership has deep roots within the SANBI community initiatives, supported by a generous dose of serendipity, and some very special personalities.

The Cape Flats Nature Partnership has been an incubator of dozens of effective, but always challenging, often frustrating and occasionally disappointing environmental projects. The anchor to all of these has been the Edith Stephens Wetland Park (ESWP) (see page 160).

Kirstenbosch ecologist, George Davis, ever restless of mind and ready for action, came upon the Edith Stephens Nature Reserve in the early 1990s. Davis was exploring options to narrow the divide between verdant Kirstenbosch and the vast sprawl of informal settlements and low-cost housing that was emerging on the denuded Cape Flats. What followed is an inspiring story of ever strengthening partnerships between Kirstenbosch, the Cape Town Metropolitan Council, community groups, NGOs and private individuals.

Today, once neglected wastelands such as Edith Stephens Nature Reserve, Wolfgat Nature Reserve, Macassar Dunes Conservation Area, Rietvlei and Harmony Flats Nature Reserve have become centres of vibrant, community-driven

The Kirstenbosch Outreach Greening project has involved learners from over 100 schools in developing hands-on skills in gardening.

The remarkable Miss Stephens

Edith Stephens (1884–1966) was a strongly independent woman, occasionally quirky in teaching method and dress, but much loved by the generations of students that she taught during her 29 years at the University of Cape Town. She was one of the three young South African students, on scholarships at Cambridge, whose infectious enthusiasm for the fynbos inspired Robert Compton to seek his future in the Cape (see page 36).

Recognition was given to Edith Stephens by the Cape Tercentenary Foundation in 1957 for her role in the conservation of the natural flora and fauna of the Cape. Stephens used the award money to purchase a small wetland on the Cape Flats, of special interest because of the community of *Isoetes capensis* found there – a quillwort, a tiny, spring onion-like member of an ancient group of 'living fossils' related to ferns. Only someone with the keen botanical interests of Edith Stephens would have noticed the humble wetland species, let alone sought its conservation. Over time, however, the small patch of wetland donated by Edith Stephens to Kirstenbosch became increasingly isolated from its mothership, adrift in the growing urbanisation, industrialisation, confusion and decay of the Cape Flats. In recent years, thanks to the efforts of Kirstenbosch ecologist George Davis, this reserve – albeit surrounded by dense urban settlement – is once more a haven for nature.

Edith Stephens (1884–1966) played a pioneering role in conserving Cape wetland ecosystems.

The Edith Stephens Wetland Park links Cape Flats communities with nature.

environmental projects, supported by the Cape Town Metropolitan Council, the SANBI, the Table Mountain Fund and many others. In their comprehensive synthesis of the partnership (*Growing Together: Thinking and Practice of Urban Nature Conservators*), authors Bridget Pitt and Therese Boulle conclude: 'These sites may seem like play parks compared to the huge pristine tracts of our national reserves ... but these hard-pressed little foot soldiers are nature's ambassadors and are critical to the future of conservation. Every year, a higher proportion of the population is born into city life. Most of these people will never set foot in a large national park ... for these individuals, the nature conservation sites represent their *only* opportunity to have contact with nature. They are also the single places that can offer the spiritual and emotional healing which only nature can provide to highly stressed and impoverished communities.'

And they offer tribute to the anchor project, the ESWP, which they describe thus:

'In the midst of this burgeoning industria lies a small jewel: a sparkling stretch of water where egrets and cormorants congregate, a wetland teeming with frogs and aquatic life, a centre buzzing with environmental education and other activities'.

Youth groups are involved in habitat restoration projects in the Cape Flats nature reserves, seen here preparing fire-breaks and leading a clean-up campaign.

Come smell the flowers 161

Willem Boshoff's installation 'Garden of Names' flags 2 000 species of threatened plants.

Interpreting the Garden

Until fairly recently, to meet the goal of educating visitors, it was deemed sufficient to place a name label – usually just the Latin binomial and the country of origin – alongside a plant in a botanical garden. Much has changed in the last decade. Educating the visitor is no longer the objective. Effective and entertaining interpretation is the new goal – encouraging, facilitating and inspiring visitors to ask questions and seek answers themselves. Education is a bonus, not an objective.

With this new philosophy as her framework, Marÿke Honig, a young botanist/horticulturist at Kirstenbosch in the 1990s, transformed the rather dreary tradition of name labels into a new approach to enliven the Garden for visitors. Her handbook *Making your garden come alive* (first published by the SABONET Project) introduced an exciting, stimulating, but unobtrusive suite of interpretation tools. It has since been translated into French, Portuguese and Spanish for use across Africa and Latin America.

Theme gardens were developed within the Garden, enabling visitors to follow, understand and remember the

Useful Plants garden

The theme garden that attracts the most interest among school learners is the Useful Plants garden, where the rich heritage of traditional plant use is presented. Old and young, whatever their backgrounds, whether locals or visitors, all marvel at the variety of medicinal, magical, food and fibre uses to which our flora has been put over the centuries. As younger generations become increasingly isolated from their cultural roots, this garden serves as a unique link to a rich but disappearing heritage.

Eucomis autumnalis, *one of the most important sources of traditional medicine, grows in the Useful Plants garden.*

'A Hard Rain's a-gonna Fall' banner conveys a strong message on the planet's environmental and social crises.

reasons for such groupings, rather than just the names of constituent plants. Honig created lively, engaging storyboards to serve as the key interpretive material for communication in the Garden. First to be launched was the 'Water-wise garden': visitors could see what is achievable in terms of structure, colour, resilience, economy and sustainability using species adapted to seasonal or long-term drought. The 'Fragrance garden' demonstrates the diversity of the scented plants that give fynbos its characteristic accents and notes. The 'Garden of extinction' tells of the threats faced by our flora, and the successes that have been achieved by Kirstenbosch's Threatened Species Programme in propagating and reintroducing species – extinct in the wild – to their former habitats. Use of graphic displays such as Willem Boshoff's 'Garden of Names' or the photo collage on the theme of Bob Dylan's 'A Hard Rain's a-gonna Fall' have been very effective in communicating messages on global environmental crises.

A century ago, Harold Pearson lamented the poverty of South African interest in and knowledge of indigenous horticulture. This was at a time when there was not a single illustrated text on the country's flora, and more expertise on growing proteas, mesems, agapanthus or gladiolus could be found in London, Paris or Berlin, than in Cape Town. We can look back with pride at the remarkable literature now available on our flora, including the unrivalled series of *Wild Flower Guides* published by the Botanical Society, the Kirstenbosch *Grow* and *Gardening* series, and the hugely successful website www.plantzafrica.com, which gives comprehensive horticultural notes on over 1 000 species at the click of a button.

The real key to interpretation, and to the excitement that it must provide, lies in the passion of the interpreter. In Ally Ashwell's words:

'There is no substitute for the warm, enthusiastic and informed welcome which first-time visitors receive from a person who knows and loves the Garden and wants to share its secrets.'

Come smell the flowers

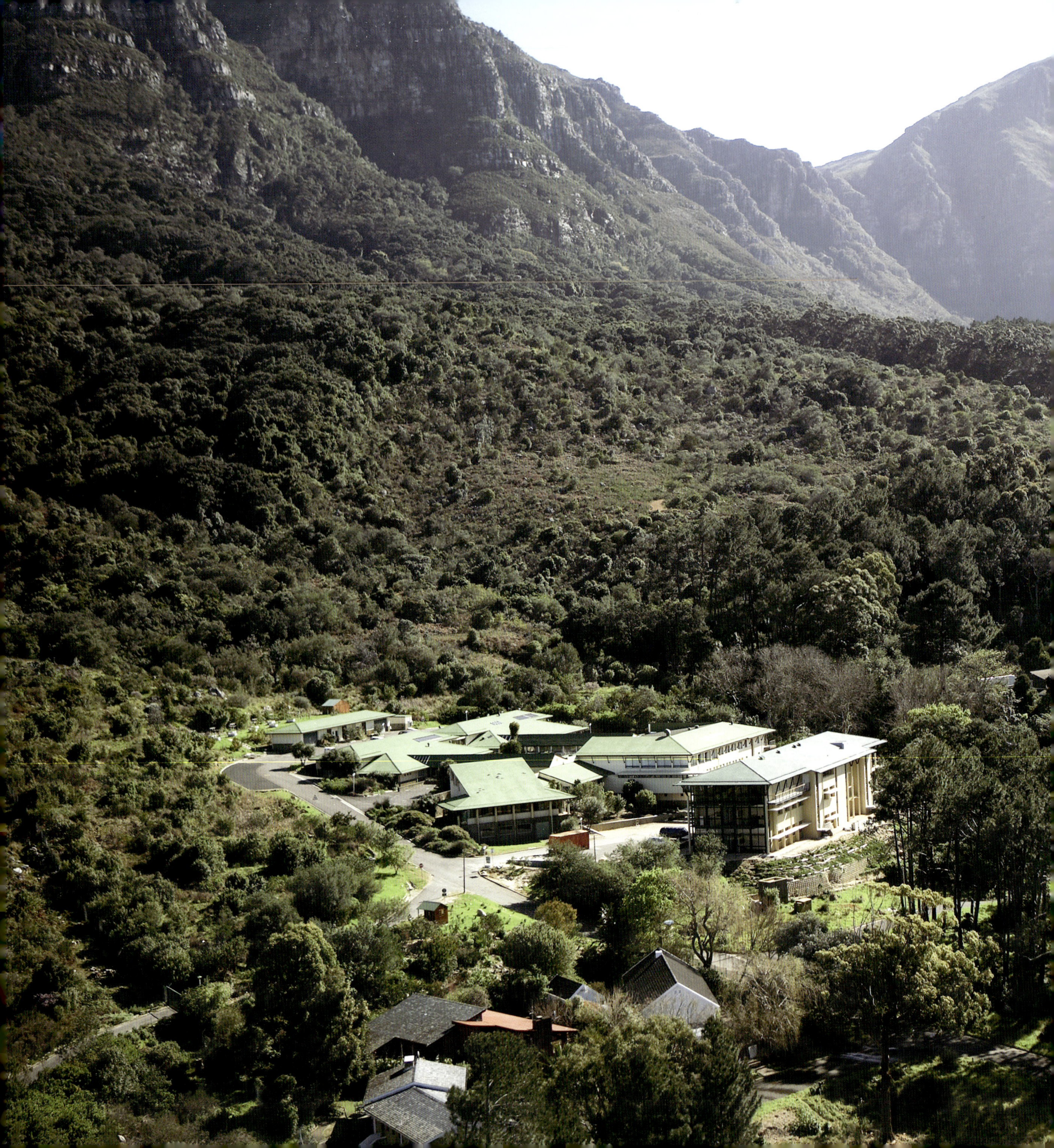

CHAPTER NINE
Conservation science

Understanding the workings of nature

Research is, or should be, undertaken with the single objective of discovering truth, regardless of the consequence. These, however, may at any time assume a practical and economic importance which no one has been less inclined to expect than the investigator himself.

HAROLD H.W. PEARSON
South African Journal of Science, *1910*

Ninety years passed before Pearson's vision of a well-equipped research facility at Kirstenbosch was realised in the Kirstenbosch Research Centre.

New directions in research

For the first 60-plus years of Kirstenbosch's history, research was a poor cousin to the main focus of the Garden – indigenous plant horticulture. The Garden's first Director, Harold Pearson, was an internationally respected scientist who dedicated his energies to the establishment of Kirstenbosch as well as practising as a full-time professor; and Robert Compton, who succeeded Pearson as Director, developed the herbarium that was to carry his name and contribute so much to our knowledge of the Cape flora. The energies of Brian Rycroft, Compton's successor, were largely channelled into expanding the network of National Botanical Gardens across the country. When Kobus Eloff was appointed to succeed Rycroft, the research team at Kirstenbosch numbered just two taxonomists. Eloff was successful in attracting external funding, and with it a team of young scientists – several of these making major contributions to developing the science programme.

But the first real change arrived in 1990, when the research strategy for the new NBI changed course from an essentially inward focus of documenting the flora of the Cape and the threats to its plant species, to addressing the emerging needs and opportunities of a rapidly changing socioeconomic landscape.

In spite of the decades of political and academic isolation, trade embargoes and social ostracism to which apartheid South Africa was exposed, international scientific co-operation continued through collegial avenues, just as it had across Europe during the Napoleonic wars, across South Africa through the South African war, and across the world during the Cold War. In the 1970s and 1980s, South African ecologists played an active role in the developing fields of fire ecology, invasive species biology, landscape change and the management of threatened habitats and endangered plants. This exposure to emerging trends in ecology and conservation led to the newly established NBI adopting a much broader sense of opportunity than previously thought important or even appropriate to the research programmes of a botanical institute. The United Nations Conference on Environment and Development, held in Rio de Janeiro in 1992, introduced new global agendas in environmental policy, management and research, with new focuses on topics such as desertification, climate change and biodiversity.

Joseph Mathews, at right, and Alec Middlemost, Assistant Curator, admire the pond in the main lawn. Mathews published numerous articles on his horticultural research into growing indigenous plants, which, with taxonomic studies led by Compton, dominated the research programme at Kirstenbosch for its first seven decades.

Environmental problems and collective solutions

Emerging from the rapidly developing international environmental consciousness of the late 20th century was the realisation that many environmental problems were soluble only through collaborative multidisciplinary research. During the years of relative isolation, South Africa had initiated many programmes to address complex environmental problems, based on the philosophy of co-operative science that embraced the efforts of universities, government agencies and the private sector. Without any dedicated budgets or specially appointed staff for these themes, from its inception in 1990, NBI tracked down funds, inspired young researchers, and developed partnerships with like-minded individuals and institutions both nationally and internationally. The new focus on collaborative research was aligned to the needs of the new, post-apartheid government.

An early project that opened new approaches to old problems was the land transformation research led by Timm Hoffman of the Kirstenbosch Research Centre (KRC). He challenged both the prevailing political system and scientific dogma in the ground-breaking book *Nature Divided – Land Degradation in South Africa*. Written with Ally Ashwell, and published by UCT Press in 2001, the book synthesised Hoffman's research on landscape change in South Africa, providing an update on the trend that Sparrman, Thunberg and Burchell had first pointed out some 200 years earlier, when they noted the impacts of overgrazing in the Karoo. *Nature Divided* remains a benchmark on the topic.

Climate change had appeared on the global science agenda during the mid-1980s, but it had been regarded as little more than a passing big science fad by some of the country's leading scientists. Fortunately, the science programme at Kirstenbosch was not constrained by the power brokers of the government's key agencies or even by the entrenched academic community. NBI's Guy Midgley

One of the first reports emerging from the Kirstenbosch Research Centre, written by Timm Hoffman and Ally Ashwell, was Nature Divided: Land Degradation in South Africa.

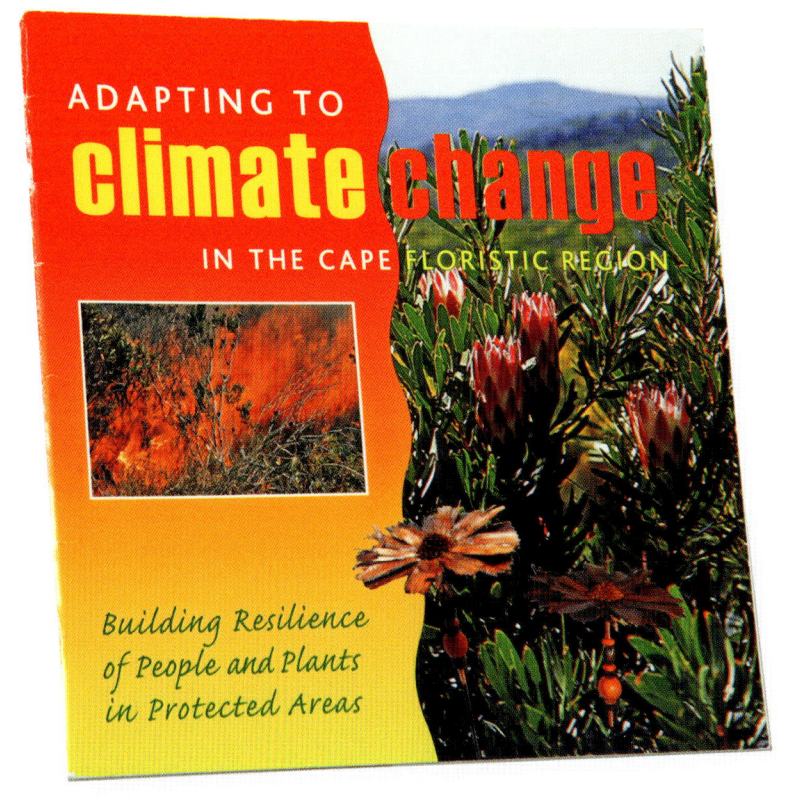

The possible impacts of climate change on South Africa's biodiversity have been a key field of study by Kirstenbosch Research Centre researchers.

Conservation science   **167**

Aloe dichotoma grows in the Nieuwoudtville Kokerboom Forest. This southernmost population of Kokerbooms is stable, but those to the north, in Namibia, are suffering from increasing rates of dieback, possibly linked to a warming climate.

and his colleagues, now established in the newly built Kirstenbosch Research Centre, were encouraged to explore, predict and explain the impacts of potential climate change on the vegetation and flora of southern Africa. The climate-change work at the KRC is one of SANBI's flagship programmes, and enjoys wide international recognition.

One of the most quoted papers emerging from the Kirstenbosch Research Centre is the study by Wendy Foden and colleagues on the distribution and dynamics of the iconic Quiver tree – *Aloe dichotoma*. Discovered and named by that indefatigable plant collector from Kew, Francis Masson, while exploring the rugged Bokkeveld in the company of Carl Peter Thunberg in 1774, *Aloe dichotoma* is found across 11 degrees of latitude (between 32 and 21 degrees south) from the arid Northern Cape through to central Namibia. Foden found that the recent dieback in Quiver trees was most pronounced in the northern populations, while southern populations were healthy. She correlated the patterns of dieback with observed and modelled climate changes in the latter half of the 20th century, and also with known temperature and aridity tolerances of the species. She concluded that the geographical range of *Aloe dichotoma* is being progressively squeezed between dieback in the north, and the slow rate, or even absence, of expansion to the south. Her findings challenged the popular belief that populations of plants will simply migrate polewards (towards cooler climates) during global-warming events. Her work furthermore suggests that desert floras, once thought to be relatively immune to climate change, will likely become more species-poor with increased global warming.

An equally important research project at the KRC, undertaken by one of Guy Midgley's Ph.D. students, Barney Kgope, and co-supervised by UCT ecologist William Bond, was on the responses of savanna trees to increased concentrations of atmospheric carbon dioxide. The rise in atmospheric carbon dioxide is well documented and shows a 25 per cent increase over the past 50 years, forming the basis of suggestions that the globe is facing a seriously accelerated warming event. What Kgope found is intriguing, and of key importance for stock farming and wildlife tourism in the savanna regions of Africa and other intertropical areas (see box opposite).

The fate of Africa's savannas?

Ph.D. student Barney Kgope undertakes field measurement of carbon dioxide uptake during photosynthesis.

Studies on the responses of savanna trees to increased concentrations of atmospheric carbon dioxide suggest that major changes in the ratios of woody plant to grass in savanna ecosystems could occur in the future.

The grasses of African savannas photosynthesise following what is called the C_4 process, which is more water efficient than the more common photosynthesis pathway, called C_3, used by temperate grasses, shrubs and trees. Growth rates of C_4 plants are only marginally influenced by the concentration of atmospheric carbon dioxide. Trees and shrubs following the C_3 process, however, are positively enhanced by increased atmospheric carbon dioxide. Using specially designed growth chambers, Kgope was able to demonstrate a marked increase in growth rate, both of shoots and roots, of savanna trees exposed to the levels of carbon dioxide expected within the next 50 years, if current trends of fossil-fuel use continue. In African savannas, the balance between grasses and trees is maintained by fire. Fires of normal frequency and heat are sufficient to kill most young trees at the sapling stage, resulting in the mosaic of trees and grassy areas typical of the African savanna. But, if saplings grow just a little faster, they can emerge above the kill-zone of fires and grow to maturity – forming a closed woodland, or worse, a dense thicket. Once a tree or shrub canopy is established, the C_4 grasses will not receive enough sunlight, their biomass will gradually decrease, fuel loads will reduce and fires will become less frequent. What was originally a productive savanna, providing a mix of grazing and browsing for cattle or wildlife populations, will transform into an unproductive thicket, with serious socioeconomic consequences.

Conservation science

Biodiversity data sets – the building blocks of predictive ecology

'Not everything that counts can be counted, and not everything that can be counted counts.' This somewhat cynical quote is attributed to Albert Einstein (1879–1955). The reality is that predictive ecology demands accurate information on the abundance and distribution of animals and plants. Much of this information on African ecosystems was laid down during two centuries of botanical collecting that preceded modern ecosystem studies. What accelerated our understanding of climate/biodiversity dynamics in South Africa was the establishment of several atlassing projects – a compendium of geo-referenced information, where the exact geographic location of the record can be fixed to an accuracy of a few metres through the use of satellite-based geographic positioning systems, or through the use of fine-scale maps. In the case of birds, South Africa has one of the finest information systems available anywhere, based on bird sightings collected by thousands of volunteers over the past 30 years. More recently, other networks have been established, making use of the power of mobile phones, digital cameras and the internet to rapidly document information on butterflies, reptiles, frogs and spiders. Many of these are now incorporated and funded through the wider science network of SANBI.

Protea Atlas Project

The Protea Atlas Project has been a Kirstenbosch initiative since its inception in September 1990. With a small in-house team, but with hundreds of volunteer field workers, Tony Rebelo has amassed what is probably the biggest, most accurate, geo-referenced database of information on the distribution and abundance of any single family of plants, anywhere. This unique database, linked as it is to detailed environmental information on climate, soils, altitude, slope and aspect, provides the statistical modeller with a mother lode of information for developing and testing ideas on the responses of species to changes in environmental factors.

Examples of the floral diversity in the family Proteaceae: ABOVE LEFT Leucadendron tinctum *and* ABOVE RIGHT Mimetes argenteus

A major thrust of the NBI's Stress Ecology Research Group (first based at the University of Cape Town as the Experimental Ecology Group, and then at the newly completed Kirstenbosch Research Centre), was the careful computerisation of the extensive plant ecological records of John Acocks, author of the classic *Veld Types of South Africa*. This invaluable data set enabled SANBI to produce a spatially clear national assessment of the vulnerability and adaptation of plant biodiversity to climate change.

Many of the recent advances in our understanding of the patterns of threat to our endemic species, and of trends and impacts of climate change, are due to the inherent value of good information on plant and animal distribution. Information management, and in particular the dissemination of biodiversity information, is now a primary objective of SANBI researchers.

John Phillip Harison Acocks (1911–1979) authored the classic study on Veld Types of South Africa, *which served the country as a benchmark for 50 years.*

ABOVE LEFT Protea cynaroides *and* ABOVE RIGHT Leucospermum erubescens

Conservation science 171

Cycads, the plants that lived with dinosaurs

Cycads, their ancient origins and their present precarious situation have attracted scientists to Kirstenbosch from the Garden's establishment to the present day. Cycads are the oldest of seed plants, having survived three of the Earth's mass extinction events, found through 280 million years of the fossil record. Globally, fewer than 400 species of cycad are still living. Perched as they are on the edge of extinction, most of South Africa's 38 species are represented in the living collections established from 1913 by Harold Pearson, nurtured by John Winter and Dickie Bowler and studied exhaustively by John Donaldson and colleagues.

John Donaldson, head of SANBI's Applied Biodiversity Research Programme based at the KRC, is a world authority on the biology and conservation of cycads. As Chairman of the IUCN Cycad Specialist Group, and Chair of the IUCN/SSC Plants Committee, he is well placed to appreciate the crisis faced by this charismatic group of living fossils. South Africa is one of the global hot spots of cycad diversity, with 68 per cent of its 38 species threatened with extinction and three of these listed as Extinct in the Wild – just a short step from oblivion.

Unlike the majority of South Africa's threatened species, which have been reduced through land transformation and habitat loss, the precarious status of cycads exists because of illegal trade. Thousands of cycads end up in the hands of private collectors, prepared to pay top prices to augment their collections. In addition, the stripping of bark for traditional medicine has led to the complete loss of populations of some species in KwaZulu-Natal and the Eastern Cape. Despite the efforts by IUCN and conservation scientists, the best efforts have been futile, serving only

Cycads are the oldest of seed plants, found through 280 million years of the fossil record. These large specimens of Encephalartos woodii, *severely damaged by bark removal, were photographed in 1918 at Ngoye, Natal, the last specimens known in the wild.*

Now extinct in the wild, Encephalartos woodii *survives in many botanical gardens, but because only male plants have ever been found, vegetative propagation from suckers is the only option for expanding the living collection.*

Despite these healthy male cones of Wood's Cycad Encephalartos woodii, *seed propagation of this species is impossible as no female plants exist.*

to document what is a botanical disaster and a national disgrace. Species such as *Encephalartos inopinus* have been monitored by conservation authorities in Limpopo since 1992. They have recorded this species' decline from 677 specimens in 1992 down to 81 in 2004 and there are unsubstantiated reports that the species might now be Extinct in the Wild. A classic case of going, going, gone.

Fortunately, SANBI has, within its network of gardens, a diverse gene bank of specimens that can be used in 'captive breeding' programmes, possibly leading to their eventual re-establishment in their former habitats. But, as John Donaldson's research has shown, cycads are dependent on symbiotic relationships with insect pollinators, algal nitrogen fixers and mammal dispersers of their seeds to maintain their populations. Restoration projects are far more complex than merely planting nursery-grown specimens back in the wild. Cycads remain the most critically threatened plant group in the world.

Most of South Africa's cycads are being driven towards extinction by illegal trade. These confiscated plants have been donated to the NBGs for use in SANBI's species recovery projects.

Conservation science 173

What had been an overgrown thicket of poplars, brambles and acacias in 1913 has been transformed into a cycad grove backed by a forest of indigenous Real Yellowwood Podocarpus latifolius, *Small-leaved Yellowwood* Podocarpus falcatus, *Mountain Cedar* Widdringtonia nodiflora *and Cape Holly* Ilex mitis.

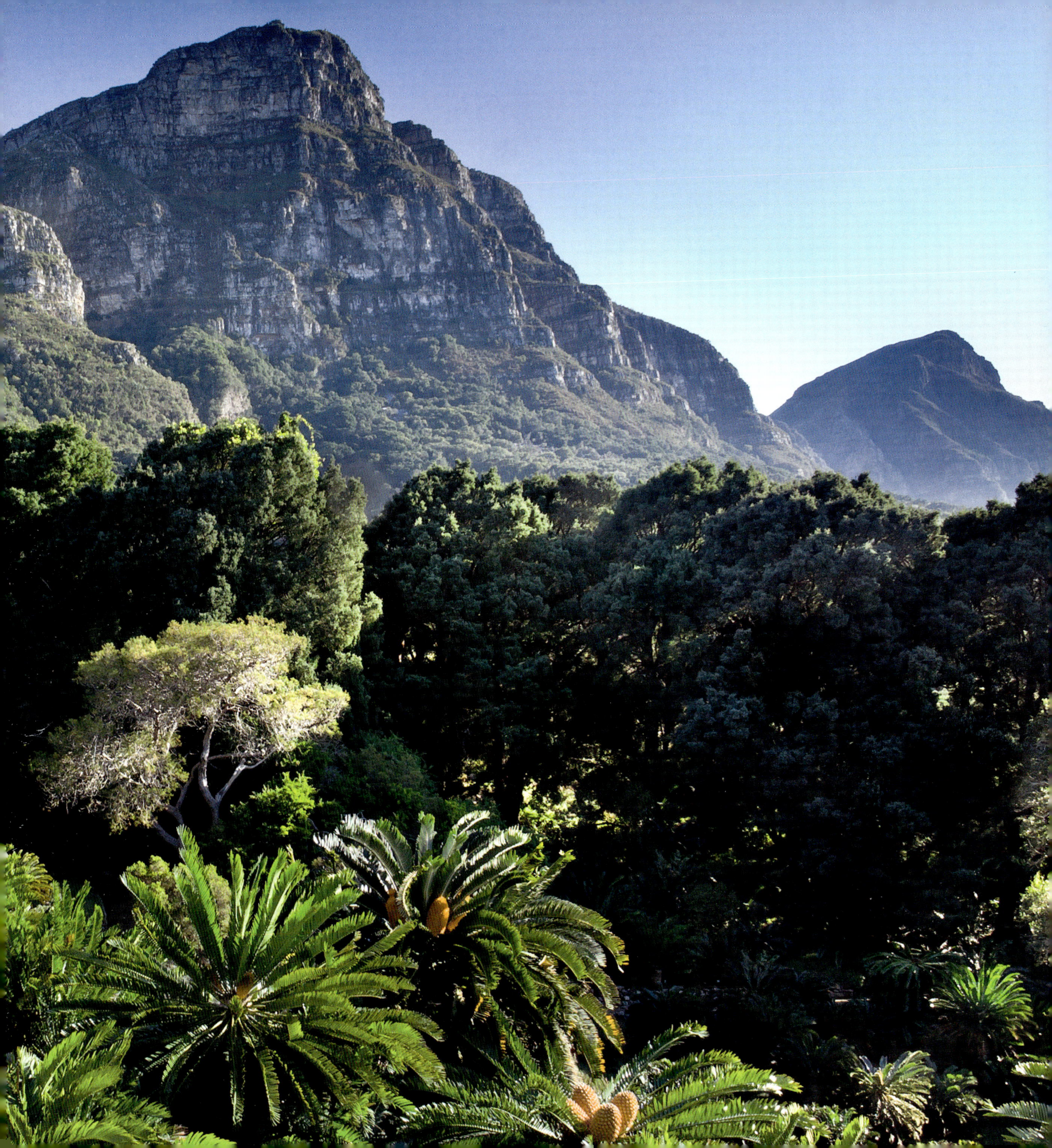

The Kirstenbosch smoke primer

Research is often a serendipitous endeavour. Lord Rutherford is quoted as saying that 'you can plan research, but you cannot plan discovery'. Such was the experience of Hannes de Lange, who in 1986 joined the Kirstenbosch Research Centre team. De Lange set about studying the life history of *Audouinia capitata*, a rare Cape endemic of the family Bruniaceae, a key fynbos family, but one that had defied propagation for use in horticulture. Careful field observations convinced De Lange that the fires that characterise fynbos ecological processes must have a role in the germination of the abundant seeds produced by fynbos species. He experimented with all possible combinations of heat intensity and frequency, different ash loads, both in the laboratory and in the field, burning patches of veld in different seasons and different fynbos communities – but nothing would induce the seeds to germinate. In desperation he wondered whether the smoke billowing forth from veld fires might play a role in the process. He set up a simple device, adapted from a smoker used to smoke trout, and subjected samples of seeds to smoke from the kindling of fynbos plants. *Eureka!* – within a few days the seeds germinated. From a near-to-zero germination rate, De Lange and his colleagues were obtaining over 90 per cent success for many fynbos species that had previously been impossible to propagate. News of his success spread like the very wildfire that was the key to his idea, and soon researchers at Kirstenbosch's sister garden at King's Park in Perth, Australia, and at the University of KwaZulu-Natal had isolated the active chemicals involved in the process. De Lange's colleague at the Kirstenbosch Research Centre, Neville Brown, produced a most practical way of exploiting the discovery. By pumping smoke through a water bath, soaking filter papers in the water and then drying them, he was able to store the active compound for future use. He packaged the filter papers and distributed them as the 'Kirstenbosch Primer'. Gardeners needed only to wet the paper in a plate or Petri dish, place seeds on the damp, smoke-primed surface, and wait for them to germinate.

A simple device, adapted from a 'smoker' for smoking trout, was used to collect the molecules emitted from veld fires for seed germination studies.

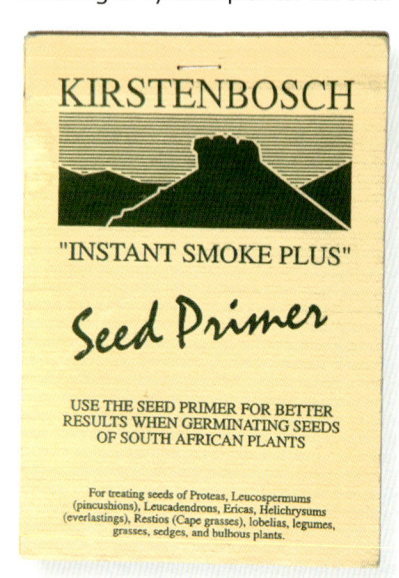

The Kirstenbosch 'seed primer' is a clever innovation for enhancing seed germination.

The Restio trial beds have proved to be a great success: the robust Cannomois grandis *seen here is excellent for landscaping.*

Into the age of molecular biology

Students prepare samples for phylogenetic studies in the Leslie Hill Molecular Laboratory.

One of the remarkable features of the research programme at Kirstenbosch is its breadth and depth. Following the wonderful philosophy of René Dubos – 'think globally, act locally' – the research agenda includes studies at the global scale (on climate-change modelling); at the continental scale (on plant taxonomy); at the national scale (biodiversity assessments, species conservation, vegetation mapping); and down to the molecular scale (studies on the evolution and relationships of selected plant and animal groups).

The establishment of a molecular laboratory at Kirstenbosch was a great advance in the modernisation of its facilities. By a stroke of good fortune, Leslie Hill, a long-time supporter and benefactor of Kirstenbosch, called one day to discuss the interest he had developed in molecular biology. Having read an article on molecular phylogenetics in the magazine *Scientific American*, Leslie, at the age of 92, expressed the view that Kirstenbosch should have the facilities to undertake such cutting-edge research. A few months later, and with his gift of R1.5 million, the Leslie Hill Molecular Laboratory was established at the KRC in 2000. The molecular biology programme has expanded from a focus on the evolution of proteas and other fynbos plants to include animal groups with special importance in South Africa, in particular reptiles and frogs. Our national obsession with the big, hairy mammals means we have tended to overlook our lizard fauna. With 267 species, of which 53 per cent are endemic, it is the third richest in the world, even richer in endemics than our terrestrial mammal fauna of 249 species, of which only 14 per cent are endemic to the country.

The use of sophisticated technologies such as those available at the Leslie Hill Molecular Laboratory allows researchers to explain often controversial phenomena. A well-documented observation in the Cape is the recent expansion of the range of the Hadeda Ibis *Bostrychia hagedash*, explained by the increased availability of nesting sites and feeding habitats in the expanding wooded suburbs of Western Cape towns over the past century. Hadedas were certainly unknown in Kirstenbosch during Pearson's time, yet they are now very much part of the Garden's avifauna. Less easily explained was the westward expansion – by 500 kilometres during the last decade – of the Painted Reed Frog *Hyperolius marmoratus*. To understand this, molecular techniques were needed.

Early hypotheses proposed that the expansion was due to climate change – a much over-stated cause of many recent environmental phenomena. But careful genetic studies by Krystal Tolley, leader of the molecular biology group at the KRC (see panel opposite), and colleagues at the University of Stellenbosch have shown that the expansion can only be due to direct 'human-mediated jump dispersal' – a fancy way of saying that frogs or their fertile eggs might have been

The Painted Reed Frog Hyperolius marmoratus *is a newcomer to the Cape, accidentally introduced by humans.*

A 'cradle of faunal diversity'

Krystal Tolley, leader of the molecular biology group at the KRC, has been undertaking some fascinating research using the Cape's rich diversity of Dwarf Chameleons as models of climatic, ecosystem and evolutionary change that has taken place over the past 14 million years. Using DNA sequence data for the 15 described species, plus a possible nine additional new species of Dwarf Chameleon, Tolley has established the correlation between speciation events (that is, the evolution of new species) and the changing environment of southern Africa. Her results indicate, furthermore, that southern Africa may truly be a 'cradle of faunal diversity' – a characteristic well established for plants, but not widely recognised for animals.

The beautifully decorated Cape Dwarf Chameleon Bradypodion pumilum *is a rare sight in the Garden.*

intentionally or accidentally transferred from their natural habitats in the east of the country to streams and dams further west. Reed frogs are often found in aquatic plants, transported around the country by nurserymen, homeowners and aquarists – so this frog is expanding its range as a result of being transported by humans, without any environmental change being needed. The noisy, entertaining and occasionally aggressive Helmeted Guineafowl and Egyptian Geese were introduced to decorate both garden and table, making their presence in the Western Cape another product of 'human-mediated jump dispersal'.

Science in service of society

As Pearson's prescient thinking repeatedly illustrates, moving from basic science to the application of its results has increasingly influenced the role that Kirstenbosch, and particularly the Kirstenbosch Research Centre (KRC), has played in the country. On global and regional scales, climate change and land-degradation studies have contributed to a predictive understanding of how our ecosystems work and how they might be managed sustainably. The fine-scale information gathered by field biologists and synthesised by modellers and spatial planners has guided the selection of sites deserving priority action for conservation. Research at the molecular level helps us to understand problems of invasive species. Gradually, the collective scientific studies and research practised under the auspices of Kirstenbosch have fed into determination of future policy and planning.

Making science relevant to everyday issues received a boost from the Fynbos Biome Project (1970s–1990s). The legacy of this project has been continued through the Fynbos Forum, where researchers, land managers and civil society discuss issues of importance to the sustainable use of fynbos ecosystems. Although Kirstenbosch researchers played no part in the Fynbos Biome Project itself, the KRC and the Centre for Biodiversity Conservation (CBC), have provided the institutional and physical home for the audaciously ambitious, and highly successful Cape Action for People and the Environment (CAPE) programme. This massive programme was conceived and initiated by the strong team of environmental scientists and managers who cut their teeth on the Fynbos Biome Project. Seizing the opportunities offered by a new political dispensation and access to unprecedented funding available from the Global Environment Facility, the team planned a far-reaching strategy for the 20-year project, launched in 2000. The history of the programme, and the contributions of the 100-plus people who made it happen, are presented in the publication *Fynbos Fynmense – people making biodiversity work*, published by SANBI in 2006.

CAPE epitomises the value of sharing resources. Kirstenbosch researchers, horticulturists, environmental educators and managers each contributed in small but meaningful ways towards the programme's success. When the full scale of the investment in the programme – over R100 million – became evident, it crystallised the need for a Centre for Biodiversity Conservation, physically linked to the KRC, as a stable, robust home for the programme. The generous support of a UK charitable trust – the Rufford Maurice Laing Foundation – provided the R12 million needed to construct the Centre for Biodiversity Conservation, which is home to CAPE, the Succulent Karoo Ecosystem Programme (SKEP), the Table Mountain Fund and Conservation International's Southern African Hotspots programme. The CBC's conference and workshop facilities have become a melting pot for debate and development of new ideas

Biodiversity for Development – South Africa's landscape approach to conserving biodiversity and promoting ecosystem resilience *is a benchmark synthesis of work resulting from several decades in co-operative conservation programmes facilitated by SANBI.*

Fynbos Fynmense – people making biodiversity work *documents the highly successful Cape Action for People and the Environment (CAPE) programme, based at the Centre for Biodiversity Conservation, Kirstenbosch.*

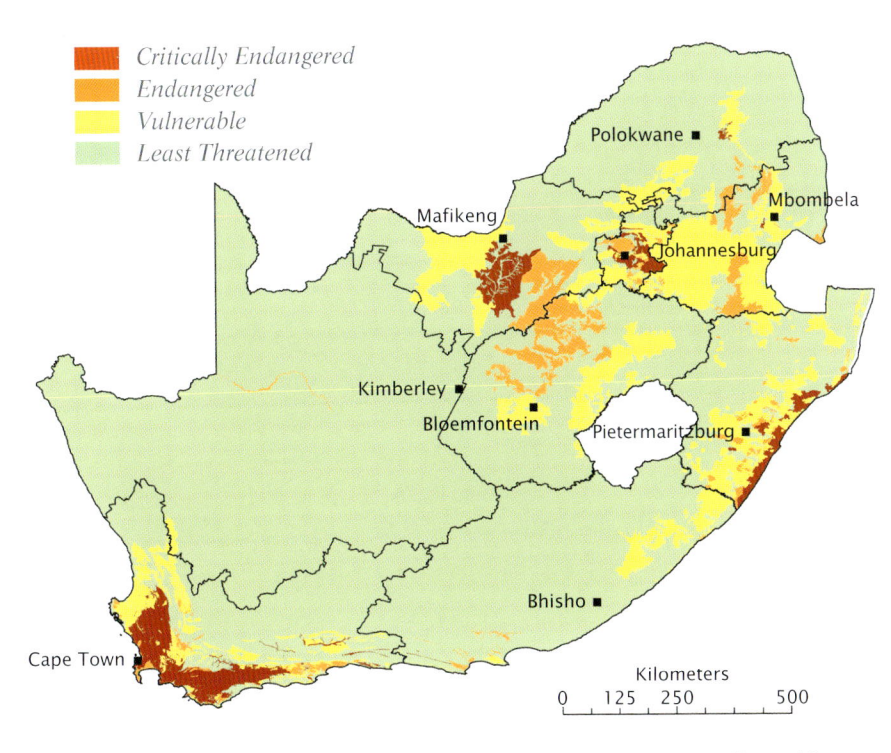

Critically Endangered
Endangered
Vulnerable
Least Threatened

South Africa's National Biodiversity Assessment 2011 was co-ordinated by SANBI's team at the CBC. The status of terrestrial ecosystems presented in this map guides decision-makers on conservation priorities.

and projects, a vibrant centre for intellectual growth, political dialogue and pragmatic action. SANBI's leadership in mainstreaming biodiversity concepts into economic planning, through its role in producing the country's first National Spatial Biodiversity Assessment in 2004 and the second in 2011, its first Protected Area Expansion Strategy, and the synthesis of lessons learned (published as a benchmark volume titled *Biodiversity for Development – South Africa's landscape approach to conserving biodiversity and promoting ecosystem resilience*), place both Kirstenbosch and the Institute at the centre of innovation and relevance.

From humble beginnings, Kirstenbosch has expanded its science programme far beyond its origins in traditional plant taxonomy, through vegetation ecology, global-change modelling and molecular biology, to the full spectrum of planning and policy processes that link science to society. In the World Bank's foreword to *Fynbos Fynmense*, the success of South African projects such as CAPE – and of Kirstenbosch, as a microcosm of the organisation – is summarised thus:

'Good scientific information and subsequent awareness raising; institutional capacity and commitment; strategic cross-sectoral co-ordination and public-private partnerships; and entrepreneurship by the conservation community in seizing the opportunities to demonstrate that good biodiversity management is good for the economy, good for local development and good for business.'

These words apply to Kirstenbosch as a microcosm of the CAPE programme – where science meets society.

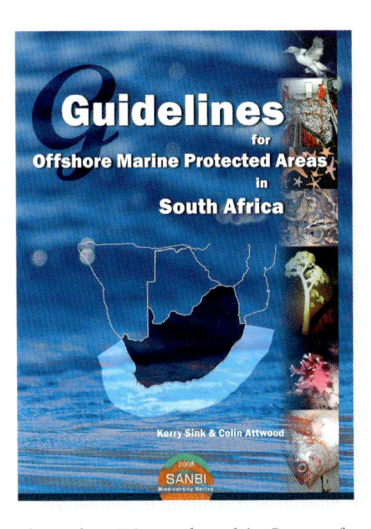

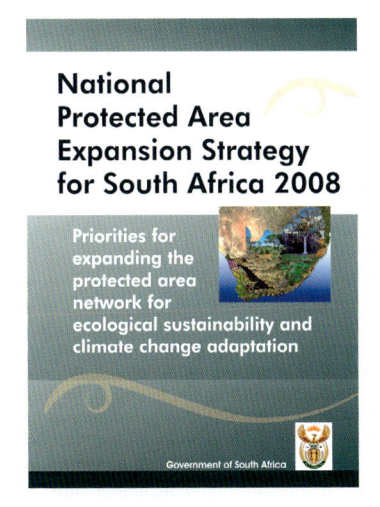

Some recent national policy guidelines co-ordinated by the SANBI team based at Kirstenbosch's Centre for Biodiversity Conservation

Conservation science

A haven of biodiversity

An interesting diversity of animal groups is to be found in the Garden, including some rare Cape endemics. Perhaps the least familiar to visitors is the enigmatic Velvet Worm or *Peripatus*, a relict from Gondwana times, unchanged from its ancestors of 530 million years ago – a rare find among forest floor logs and leaf litter. Less elusive are the Garden's butterflies. Pioneer fynbos ecologist Rudolf Marloth first described the pollination by the Mountain Pride Butterfly of the red flowering ground orchid *Disa uniflora* – the provincial floral emblem of the Western Cape. An exceptional sight is the Critically Endangered Table Mountain Ghost Frog, known only from a few forested streams in the gorges of Table Mountain. More easily seen is the Marsh Terrapin, found in the large pond in the centre of the main lawn. Snakes are seldom seen in the Garden, but the Boomslang is occasionally encountered. Although highly poisonous, this back-fanged snake is an effective predator of birds, but avoids humans. While most mammals have disappeared from Table Mountain, the Cape Grysbok, a shy but resilient endemic to the Fynbos Biome, is still resident in Kirstenbosch. Few mammalian predators survive on the Peninsula: Caracal and the Cape Fox are occasionally spotted in Kirstenbosch, while the more common and inquisitive Small Grey Mongoose, popularised by Ali Corbett's stories of *Monty the Mongoose*, is a favourite among younger visitors.

A sampling of Kirstenbosch wildlife: CLOCKWISE FROM TOP RIGHT *Velvet Worm, Marsh Terrapin, Cape Fox, Small Grey Mongoose*

182 KIRSTENBOSCH

CLOCKWISE FROM TOP LEFT *Mountain Pride Butterfly, Cape Rain Spider, Boomslang, Southern Rock Agama, Table Mountain Ghost Frog, Cape Grysbok*

Conservation science 183

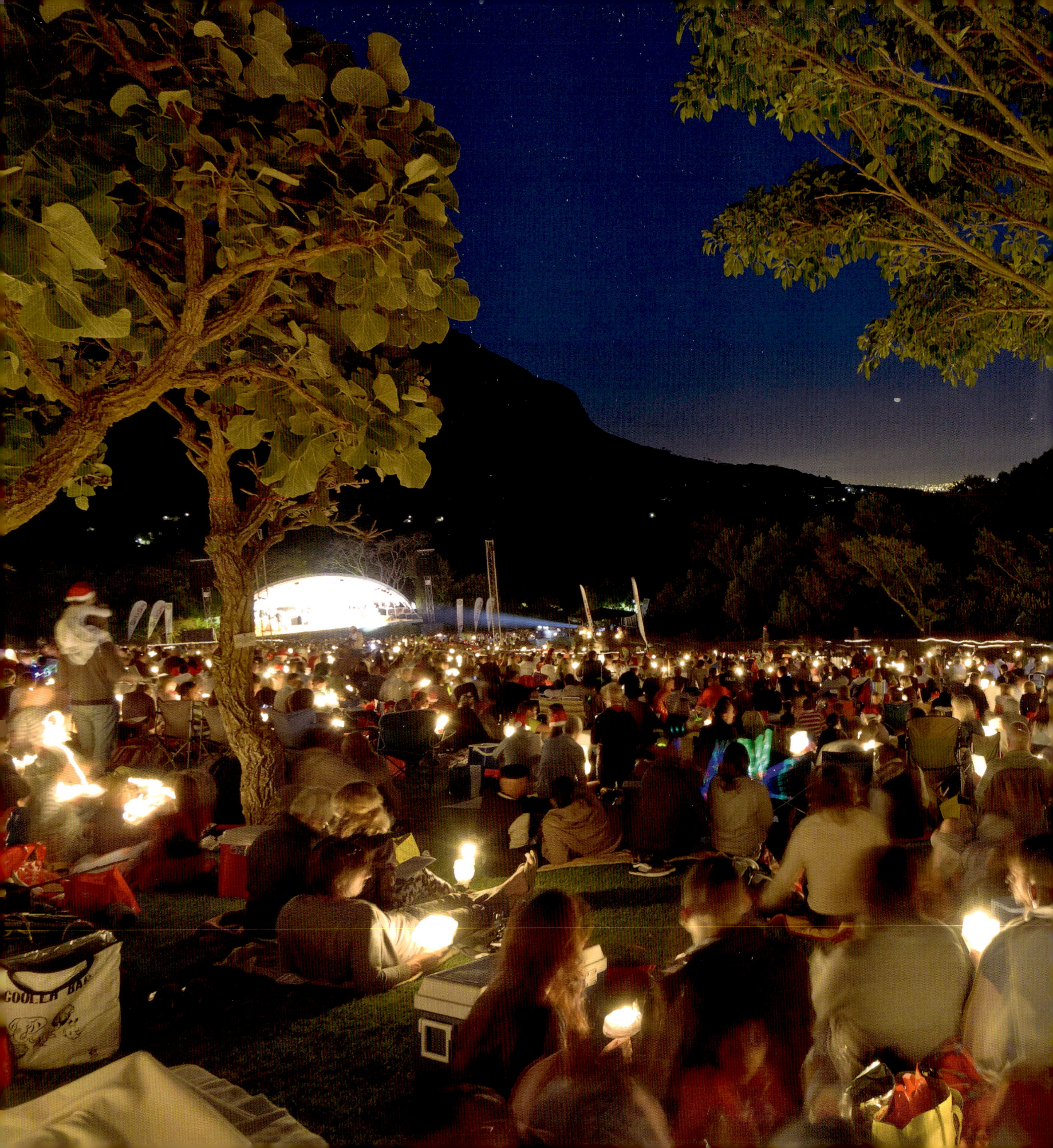

CHAPTER TEN

From poverty to prosperity

Achieving financial sustainability

South African botany has to carry on a perpetual fight against insolvency.

PETER MACOWAN
Director of the Cape Town Municipal Garden, 1881

Carols by Candlelight – a hugely popular traditional family event – is presented every December by the Kirstenbosch Rotary Club, with substantial profits going to local charities.

Early struggles

One of the valuable legacies of our colonial past is the tradition of detailed annual reports demanded of the officials of colonial governments across the globe. The history of Kirstenbosch is awash with the tears of cash-strapped directors.

The total budget for 1914 – the first full year of operation of Kirstenbosch – was £2 285. The government grant was £1 000, the Botanical Society contributed £432, and Cape Town City Council, £300. But the Garden was kept going through the sale of firewood (£209) and acorns, sold as feed for pigs (£283). In 1917, income from firewood and acorn sales exceeded that of both government and Botanical Society grants. It is truly amazing that the Garden survived its early, formative years and indeed undertook impressive developments.

Regular commissions and committees were mandated by the government to review the state of affairs and financial needs of the Garden. Each of these recommended substantial increases to the Kirstenbosch grant – but to no avail. By 1979, the Garden was still in 'a critical financial situation'.

Funding for major infrastructure was never adequate. The first government-funded project in the Garden – the construction of the director's residence, costing £2 500 – occurred in 1915. More than 50 years would pass before a project of similar cost was executed. In the interim, many small buildings appeared across the Kirstenbosch landscape. Essential needs were met in a rather ad hoc manner. Importantly, first attention was always given to basic horticultural needs. Glasshouses, workshops, staff accommodation and storerooms appeared in surges of activity at approximately 10-year intervals. In 1919 three small workers' cottages were built – the charming 'Stone Cottages' below Rhodes Drive. A curator's house was built in 1929. In 1937 a stone Gatehouse was added alongside the elegant new entrance gates and Bell Tower, and in 1939 a Lecture Hall was built. In 1947 the curator was provided with a new house, and in 1957 the Botanical Society head office was built. But, until 1969, all the buildings of consequence had been funded by public donations. In 1923, building of the Bolus Herbarium (later the Compton Herbarium) was financed through the Bolus bequest. The original Tea House was built through funds raised by the Botanical Society in 1924 and the Pearson Memorial Hostel was built in 1925, primarily with funds raised by the Botanical Society.

Buildings were erected wherever it seemed convenient – there was no master plan. As Professor Compton remarked 'Mistakes were made, naturally enough, showing themselves and having to be corrected later…' Lack of regular, predictable funding from government meant that the Kirstenbosch management team had to grab whatever funds were offered. Often the planning was done in far-off Pretoria, following standard Public Works Department specifications, with unfortunate results. The first major government-funded project since the director's residence (of 1915) was a new

A long history of financial woes

The annual reports generated by Kirstenbosch make for entertaining reading. Successive curators presented reports in impressive detail – wonderful chronicles on the trials and tribulations of the early years of the Garden.
A few extracts indicate the financial pressures of the times –
1915 Annual Report – 'the government grant reduced …'
1932 Annual Report – 'owing to financial stringency, personnel reduced …'
1946 Annual Report – 'staff and labour difficulties increased …'
1951 Annual Report – 'financial situation remains uncertain …'
1955 Annual Report – 'a stranglehold has been placed on the development of the garden …'
1979 Annual Report – 'a critical financial situation …'

RIGHT *By the 1970s, many small buildings had appeared across the Kirstenbosch landscape, as indicated in this map from the detailed* Kirstenbosch Development Study *(1973), funded by the Botanical Society and presented to the Board by architect Gabriel Fagan and his team of consultants.*

BELOW *The charming 'Stone Cottages', built as accommodation for Garden staff in 1919, now provide office accommodation for the Kirstenbosch branch of the Botanical Society.*

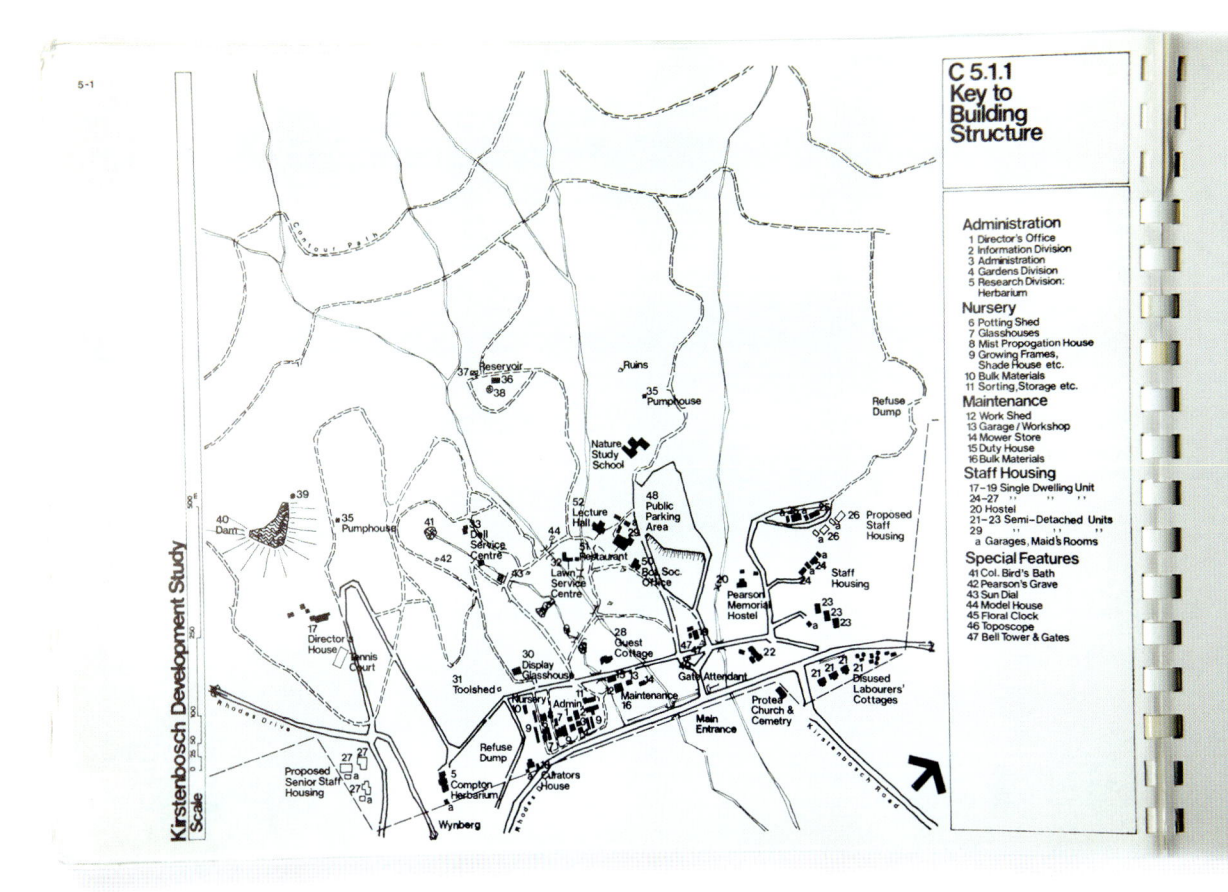

From poverty to prosperity 187

Nature Study School built in 1969. Placed high above an extensive new parking area, and of a 'Bauhaus' design, the building for many years protruded as a square concrete bunker. Careful planting of trees and sensitive landscaping now hides it. Given time, nature allows mistakes to be 'corrected later'.

The difficult years of financial constraint continued until the 1990s. What changed Kirstenbosch's fortunes was not an influx of government funding, or a windfall resulting from the sweeping sociopolitical changes of the 1990s, but rather a change in the perception of what a modern botanical garden should be. It required a paradigm shift in management model.

The Pearson Memorial Hostel was originally built to accommodate 'lady horticulturists'. From 1990 to 2004 it served as offices for the NBI senior management and the Board.

A new management model

By the 1980s, Kirstenbosch was already a world-famous and stunningly beautiful garden. It had won international acclaim from the botanical fraternity, delegations of which arrived to celebrate its Golden Jubilee in 1963, and its 75th Anniversary in 1988. But it was still a garden for gardeners and botanists; it had yet to serve society as a whole effectively. Equally, it was still wholly dependent on government funding for both day-to-day operations and capital development funding. It was hobbled by the belief that 'Pretoria would provide'.

In looking for advice from colleagues around the world, Huntley, the new Director from 1990, was surprised to find that all the famous botanical gardens – from Kew to New York to Sydney to Singapore – suffered from the same malaise: a sense of entitlement seemed to prevail, a legacy of the colonial era where botanical gardens were as much pleasure parks for the privileged, as they were centres of botanic research. Most were isolated from the local economy and few considered financial independence as an achievable goal. But a hint of what to do came from an unexpected source – Barry Ambrose, Director of the Royal Horticultural Society Enterprises at the Society's Wisley Garden outside London. Ambrose visited Kirstenbosch in 1992 to advise on planning the new garden shop. He made an interesting observation on what a successful garden required:

'All you need to succeed is good parking, clean rest rooms, a tea room and a friendly face at the gate'.

Old and new parking areas – from dusty arrival to a shady welcome. Adequate and secure parking is the first requisite of any tourist destination.

Ambrose's message was about 'user friendliness', as it is now termed. The Kirstenbosch Development Campaign was focused on providing the Garden with essential infrastructure. But it needed more than this: it needed to change the image and the user profile of the Garden. As any business person knows, the more diverse one's products are, the more customers one can attract; and the higher the quality of one's services, the better chance one has of succeeding. This is the philosophy that Kirstenbosch has followed from 1990 to the present day.

The Campaign's success was due to a new approach, one of 'strategic opportunism': strategic in that it was guided by a framework of policies and priorities; opportunistic in that it depended on lobbying, good luck and good timing for its fund raising. The consequent physical development of Kirstenbosch provided a platform for the real transformation programme – a web of strategic partnerships.

The power of partnerships

On 27 May 1913, when General Louis Botha, as Prime Minister, approved the establishment of Kirstenbosch, he had the good foresight to comment, regarding the government grant:

> *'... it is only the assurance that the sum will probably be augmented from other sources that has induced the Government to agree to the proposal'.*

Here he was referring to a new, more public support group for the Garden – one that took no time for the great and the good of Cape Town to establish. Within days, on 10 June 1913, a large public meeting chaired by the mayor of Cape Town resolved to establish a National Botanical Society. The government, too, was quick to respond – Kirstenbosch Estate was transferred to the Board of Trustees on 1 July 1913. The National Botanic Gardens of South Africa had been established.

Among the objects of the Botanical Society, one is of particular relevance to the Garden's success: 'To augment the government grants towards developing, improving and maintaining fully equipped botanical gardens, laboratories, experimental gardens, etc., at Kirstenbosch'.

Loyal benefactors Mary Mullins, Leslie Hill, Kay Bergh and Julian Ogilvie Thompson ensured the success of the Kirstenbosch Development Campaign.

Generous funding was used to build the Rufford Maurice Laing Centre for Biodiversity Conservation. The Centre has become a vibrant hub of conservation research, planning and policy development in southern Africa.

We have already seen how the 'BotSoc', as it is affectionately known, played a critical role in raising funds for both general operational costs and several major projects during the Garden's formative years. Faced with a challenging 'shopping list' of projects to champion, the Kirstenbosch branch of the Society led the Kirstenbosch Development Campaign fund-raising drive from 1990 to 2000, mobilised by its dynamic chairperson, Kay Bergh. The full story of the Campaign has been described in chapter 5 (see page 102), but it is worth recounting a few anecdotes of how 'good timing and good luck' contributed to its success.

From poverty to prosperity

Early on in the planning of the Campaign, it was noted that one benefactor to the Garden had left a cheque for R50 000 at the Garden office, as an unspecified donation. When approached over a cup of tea as to how she would like to see the money allocated, Mary Mullins replied 'Anything that is good for the Garden. And I like round numbers, let me give you another R50 000'. By the end of her 90-plus years' association with the Garden she had donated over R1 750 000 to the Campaign!

In early 1992, another friend of the Garden, Alec McGregor, called in one morning in response to an article in the Society's magazine *Veld & Flora*, in which a request for funding for the Garden's book collection had been made. In discussion, Alec agreed to sponsor a computer, but asked if there might be something else needed. When it was mentioned that Kirstenbosch really needed a library, he replied 'We do libraries'. Within a month, the Molteno Brothers' Trust, of which McGregor was Chair, committed nearly R1 million towards a library in the new Kirstenbosch Research Centre.

Such has been the pattern of support that the Garden has enjoyed from those who know it and love it. But the web of friends needed to be expanded.

A natural amphitheatre

Initially, Kirstenbosch concerts were free. At the Botanical Society's suggestion, a 'silver collection' was introduced in the second season. As the concerts' popularity rapidly grew, the cost of hosting major performers became too high, and sponsors were sought. For 15 seasons the fruit-juice company Appletiser sponsored the series, as well as the building of a concert stage in the new venue – a natural amphitheatre with splendid views of the Garden's mountain backdrop and across the Cape Flats to the distant Hottentots Holland Mountains. Happily, the sponsorship tradition has been continued by long-time Kirstenbosch supporter Old Mutual. Income generated from concerts covers one third of all costs of operating the Garden – a key contribution to the financial sustainability of Kirstenbosch.

New ventures and risk taking

While building operations proceeded in many corners of the Garden, the heart of the Garden remained undisturbed. The little amphitheatre below the Tea House provided a perfect venue for small musical performances on quiet Sunday evenings. Sue Ross, then Marketing Manager for the Garden, invited local baroque quartets, string ensembles and the occasional vocalist to perform. Initial concerns that regular visitors to the Garden would object to the intrusion of concerts – however discreet – proved unwarranted. Audiences grew from a few dozen curious passers-by, to a hundred, to several hundreds – too many for the amphitheatre. The central lawn became the new venue; but when the Cape Town Symphony Orchestra appeared on 28 March 1993, an overwhelming 8 000 lovers of classical music arrived. This was the beginning of the 'Kirstenbosch Summer Sunset Concerts' (see pages 200–201), which now bring over 100 000 visitors to the Garden every season.

Named sponsorships had been taboo in the Gardens until the 1990s. But the Board of Trustees of the time was forward thinking and perhaps a little less conservative in its outlook than its predecessors. Pressure to offer discreet naming rights came less from the sponsors than from management. The value of sustained mutual benefit seemed obvious. This has proven to be the case. Today there is hardly a major project in the Garden that has not resulted from a partnership. And a diversity of partners brings with it a diversity of new friends, supporters and active users of the Garden.

The sculpture garden blends culture and nature.

Botanical garden or theme park?

When developing attractions to draw new visitors to a botanical garden, one has to draw a careful line between features that fit with the 'spirit of place' and crass commercialisation. For some friends of Kirstenbosch, the display of a major collection of Zimbabwean stone sculptures, presented in 1997, was already pushing the bounds. But to many, especially African visitors coming to the Garden for the first time, it provided a tangible link to traditional cultures and belief systems. The exhibition featured great artists such as Bernard Matemera, Dominic Benhura, Agnes Nyanhongo, Colleen Madamombe and Arthur Fata. It proved so popular at the Botanic Gardens Conservation International Conference held in the Garden in September 1998 that many visiting directors of other great gardens – Kew, Missouri, Chicago, Sydney, Singapore, Frankfurt – immediately commissioned the collection's Curator, Roy Guthrie, to exhibit at their gardens too. The collection is still on a continuing world tour.

With a commission of 15 per cent of all sales to be paid to the Garden's Development Fund, the positive results of this partnership have been massive. A new 'sculpture garden' with all-access pathways, information booklets, cultural events and training workshops has brought tremendous credit and a significant flow of funds to the Garden.

Hosting an exhibition of Zimbabwean stone sculptures introduced a successful approach to linking cultural traditions to the natural beauty of the Garden.

TOP *Bernard Matemera's 'Blind man'*

ABOVE *Gideon Nyanhongo's 'Elder reflects'*

LEFT *Sylvester Mubayi's 'The Bira'*

From poverty to prosperity

Select works of art

Over the years, only a few select works of art have been accepted and placed on permanent display in Kirstenbosch – two of which are the striking 'Wild Seedpod' of Arthur Fata, which is a focal piece in the Visitors' Centre courtyard, and a massive work by Bernard Matemera, 'Speak less, see more' – at the Garden Centre entrance. Another permanent installation is the family group of gorillas – hewn from fallen Stone Pines from Table Mountain by Cape Town sculptor Sam Allerton. Sensitively sited within the 'rainforest' of the Dell, the gorillas tell the story of Africa's threatened species and habitats.

Much debate attended the question of whether to permit weddings in the Garden. In the end, discretion won the day: weddings are permitted, but only if the bridal party uses a private pathway to access the Gazebo, well hidden in a romantically beautiful corner of the Garden, and on condition that they use the conference centre or restaurant for the wedding reception. The advantage of being able to offer both kosher and halaal kitchen facilities has proven particularly popular among both Jewish and Muslim communities.

A development of major value to expanding the Garden's outreach has been the Conference and Exhibition Centre, co-sponsored by the Old Mutual, a consistent supporter of Kirstenbosch. Located within the Visitors' Centre complex, the conference facilities have been used almost

The striking 'Wild Seedpod' of Arthur Fata was donated to Kirstenbosch by the Anglo American and De Beers Chairman's Fund on the opening of the new Visitors' Centre.

South African sculptor Dylan Lewis has presented several of his evocative works in the Garden.

192 KIRSTENBOSCH

A living wall encloses a gallery for a temporary exhibition of sculpture by Dylan Lewis.

continuously since their inauguration in 1998. Of special value has been their use by embassies, government ministries, multinational corporations and other influential organisations – bringing to the Garden the power brokers of both society and the economy. Many of the country's Cabinet Ministers have visited the Garden, initially in the course of their official duties rather than as gardeners or nature lovers, but have later returned with their families and friends. The name Kirstenbosch has gradually permeated all sectors of society – and, with it, the association of peace, beauty and pride.

Far from becoming a theme park, Kirstenbosch has diversified its interest and enriched its visitor experience, resulting in increased visitor numbers and increased visitor spend.

The Old Mutual Conference Centre is an ideal facility for weddings and corporate functions.

From poverty to prosperity

ABOVE *The Marquee Lawn provides an ideal setting for thematic garden parties.*

BELOW *The Moyo Kirstenbosch Restaurant sits at the foot of the Garden, close to visitor parking.*

From good friends to good finances

By 1990 the Garden was receiving some 400 000 visitors per year, but its operational costs were dependent on an 86 per cent subsidy from government. How to increase visitor numbers was one challenge; as important was how to keep them coming, and how to entice them to spend more on each visit.

As a government-supported public entity, the Garden must be potentially accessible to all South Africans, offering reasonably priced facilities and services. The profit motive should focus, by contrast, on activities that are not basic to one's enjoyment of the Garden. Special entry fees are provided for school-goers, pensioners and special-interest groups, and members of the Botanical Society enjoy free entry. Many facilities such as the Garden shops, coffee bar and exhibition hall are free of admission charges. Kirstenbosch is accessible to all.

A visitor's experience of a botanical garden must obviously start with the botanical spectacle and the Garden's setting – both natural and man-made. Repeated surveys of visitors' comments have indicated that it is the peace, tranquillity, natural beauty and 'the scent of fynbos' that they remember most. But these can form the backdrop to many other experiences, which can both entertain the visitor and provide a flow of income to the Garden.

194 KIRSTENBOSCH

ABOVE *The most ambitious project ever undertaken by the Botanical Society was the funding, with the support of many friends of the Garden, of the Conservatory. The roof of the Conservatory was designed to provide maximum cross and vertical ventilation – providing a 'cool house' rather than the conventional hothouse.*

LEFT *Visitor numbers nearly doubled between 1990 and 2010. Kirstenbosch remains one of South Africa's premier tourist destinations.*

From poverty to prosperity

Sweeping lawns, placid water features and a rugged mountain backdrop provide harmony and tranquillity to the lower garden. Visitor surveys indicate that the most cherished features of Kirstenbosch are the peace, safety and unique sense of place that the Garden evokes.

Adam Harrower

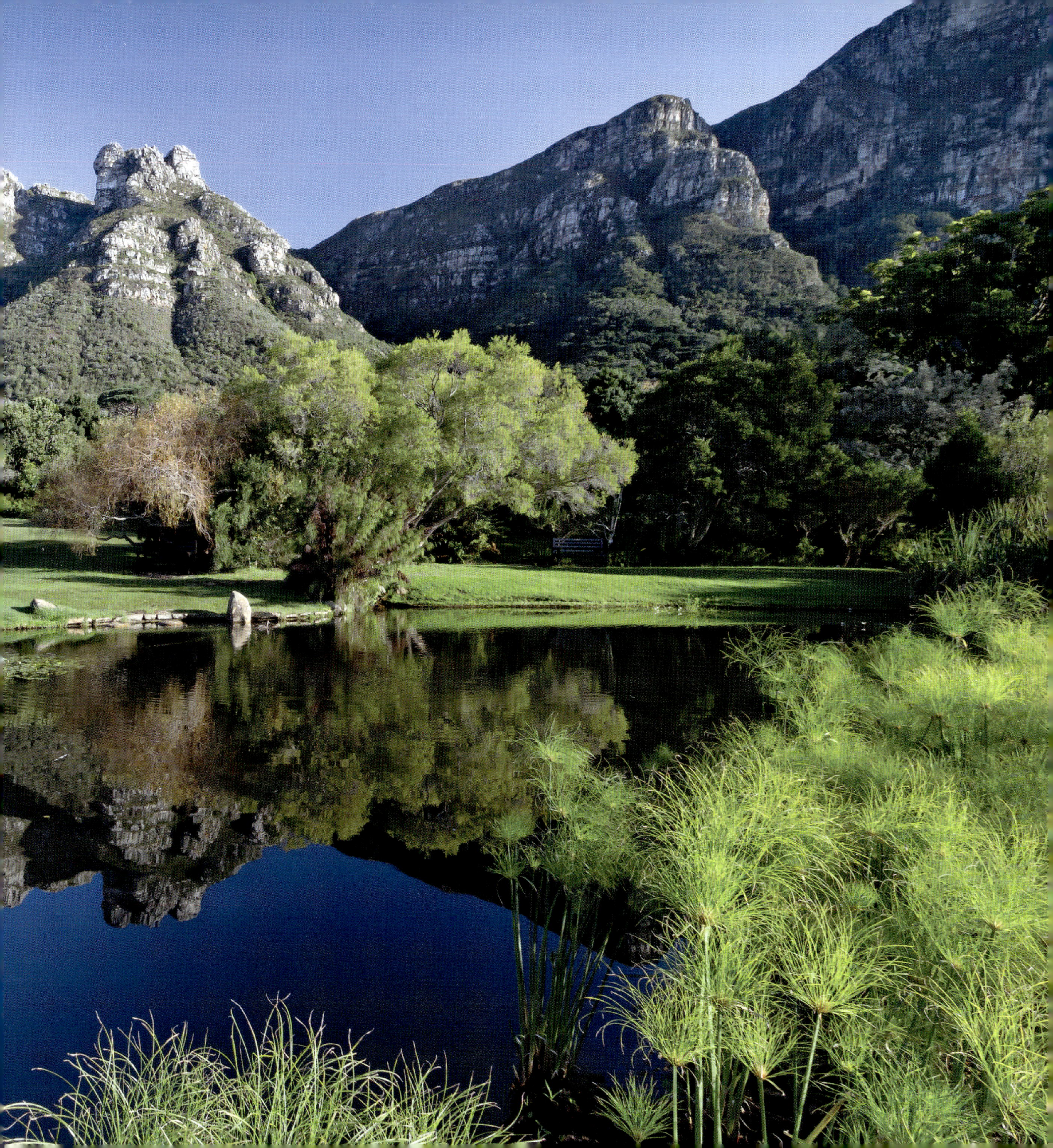

The Visitors' Centre piazza is the main arrivals point in the Garden, where shops, Conference Centre and the Conservatory are also located.

A ripple of water feeds the pool running between the piazza and Conservatory.

From 1990 to 2010, visitor numbers increased from 400 000 to 750 000 per year. Annual income (admission fees, rentals, plant sales) increased from half a million in 1990 to R30.6 million per year in 2010. Garden expenses (excluding research and education, which are separately funded) increased during the same period from R2.4 million to R28 million. Garden operations were subsidised by government to the tune of 86 per cent in 1990, but by 2010 were producing a meaningful profit – on a much larger financial base. The most striking change was in the expenditure per individual visitor, which increased from R2.75 per visitor in 1990 to R99.90 in 2010. Thus both visitor numbers and spending per visitor increased at a rate in excess of inflation, with very positive consequences.

Since 2005, the Garden has been operating at a profit, without any government contribution. Kirstenbosch is very probably one of the only large public gardens in the world to be financially independent.

Key among the processes that have led to this positive situation was the development of visitor amenities and garden infrastructure, and the promotion of a user-friendly image for Kirstenbosch as a destination for local families, foreign tourists and high-level delegations. Foreign visitors almost always include Kirstenbosch in their itineraries, with dignitaries often accompanied by senior South African politicians, such as Oprah Winfrey in the company of Nelson Mandela, and Al Gore with Thabo Mbeki.

Positive financial growth

The major sources of the Garden's income make for interesting reading. In the 2010/2011 year, admission fees contributed R13 800 000; concerts R10 100 000; tea room, restaurant and coffee bar R1 780 000; book shop and gift shop R1 680 000; hire of conference venues R1 280 000; seed and plant sales R634 000; and sculpture sales R450 000. All of these income streams continue to show steady and positive growth.

Indigenous plant horticulture has been given a boost by the Garden Centre, maintaining Kirstenbosch's tradition of service to home gardeners.

From poverty to prosperity 199

OLD MUTU
KIRSTENBOS
Every Sunday @ 17:
summer sunset concerts
Appletiser
Kirstenbosch
every Sunday
Kirstenbosch
Summer Sunset
concert
8783
Kirstenbosch
summer
sunset
concerts
This Sunday
17h30
KIRSTENBOSCH
Summer Sunset
Concerts

Kirstenbosch Summer Sunset Concerts, initiated by Sue Ross, have provided family entertainment and an unforgettable musical experience for over two million Garden visitors since their inception in 1993.

Other famous visitors who have enjoyed the Garden have been Margaret Thatcher, the Dalai Lama, Prince Charles and the Duchess of Cornwall, Princess Anne, and the kings and queens of Norway and Spain. The Summer Sunset Concerts have featured international celebrities such as Yehudi Menuhin, Elton John, Bryan Adams, Josh Groban, Cliff Richard, plus local stars such as Johnny Clegg, Freshlyground, Mango Groove and the Soweto String Quartet, attracting a wide diversity of music lovers, both young and old.

Surprisingly, Kirstenbosch has invested very little in direct marketing, using partnerships with its many sponsors to ensure that radio, press and television exposure is achieved at

The old Tea House (TOP), built in 1922, burnt down in 1988. It has since been replaced by a new Tea Room, integrated within the Garden Centre.

The Garden's partner since its founding has been the Botanical Society of South Africa, with headquarters in offices near the Garden Centre entrance.

no cost to the Garden. Many horticulturists and researchers, who initially resisted having to raise funds, became highly successful in the role. Corporate fund-raising and pressuring government for a greater slice of the national budget remain the responsibility of the CEO, supported by senior managers, to cover SANBI's ever increasing corporate needs.

Since the conclusion of the Kirstenbosch Development Campaign in 2005, few new capital projects have been undertaken in the Garden. Emphasis has been given by current Curator Philip le Roux to consolidating the financial management of the Garden's operations, while the horticultural team have focused on improving the quality of displays and living collections.

The first decade of the new millennium has seen a stabilising of visitor numbers at around 750 000 per annum, but with a steady increase in visitor spend. Operational costs have been contained by outsourcing non-core activities to service providers. Today, the Garden personnel totals 115 full-time staff, down from 160 in the 1990s. Non-core functions such as gate admission control, security and cleaning, are outsourced to service providers employing 22 contract staff. Many of the visitor amenities – restaurants, shops, tourist office, sculpture sales – employ a further 130 full-time staff. So, while Kirstenbosch has been able to reduce its staff complement, the overall employment opportunities provided by the Garden have almost doubled. The Garden is now more fully integrated into the local economy than ever – no surprise that it has received several national awards for its outstanding productivity.

Lessons learned

Pearson, in his famous 1910 Presidential Address to his learned colleagues of the South African Association for the Advancement of Science, proposed that a National Botanic Garden should engage in studies on the economic value of our flora that would 'undoubtedly do much to remove what is at present a national reproach as well as the neglect of what might be an important commercial asset'.

From its first days, Kirstenbosch was set up to garner financial reward from our flora. Pearson planted out the 'economic botany' section – most especially of buchu – and equipped it with a distillery to produce buchu essence for the overseas market. Compton struggled on with the project, finally terminating it in 1939, 'the economic idea that failed' as he later described it. During Eloff's tenure, plant production was one of the four main objectives of the NBG, but because of resistance from the commercial horticulture industry – which viewed Kirstenbosch's involvement as unfair competition – and inadequate funding for the project, this venture, too, failed. In Huntley's tenure, attempts to select and breed new cultivars for the international market through a partnership with Ball Horticulture of Chicago led to a storm of protest and emotional remarks, including accusations that Kirstenbosch was 'selling the nation's floral silver'. Ten years and many million dollars in investment later, Ball concluded that the jewels of South Africa's indigenous flora had long since been plucked for the international horticultural industry, way back in the 18th and 19th centuries, during the heyday of botanical exploration at the Cape. The hard lesson learnt is that botanical gardens should not try to engage in commercial horticultural activities. These are best left to the private sector.

But there were many positive lessons from successful initiatives and approaches. First of these, for a public service institution, was one of image. The 'friendly face' must not only have a smile – it must be professionally competent, dedicated and willing to go the extra mile. Kirstenbosch has been fortunate in having loyal, passionate and committed staff. Several staff members are third-generation employees, with many retiring after 40 or more years of excellent service.

A second key to Kirstenbosch's success has been to win friends and cherish them. The Kirstenbosch Development Campaign concluded: 'Building lasting friendships was the single most important strategy that we have followed'.

A third lesson came from the newly established democracy that transferred decision taking from a minority government to the people. Aligning strategy to government policy meant, in effect, 'keeping close to the people'. This was possible through working with the Parliamentary Portfolio Committee – which guides decisions at the highest levels of government. Hosting meetings of influential political actors has given them regular exposure to the Garden's work.

Finally, the lesson that surprised us all was that a garden can reach financial independence if all its income-generating potential is carefully developed and nurtured. The turnaround in the financial and management models introduced by Huntley and Kirstenbosch Curator Philip le Roux in the 1990s places Kirstenbosch in the enviable position of being able to look beyond mere survival – the challenge that its directors faced for the first 75 years of its history. Long may this happy situation continue!

In an oblique aerial photograph of Kirstenbosch, the Visitors' Centre, Conservatory, Millennium Glasshouses, Workshops and Garden Administration run to the left, parallel to Rhodes Drive, while the Garden Centre and restaurant lie on the mid-slopes above the main entrance off Rhodes Drive. (Photo courtesy of The Aerial Perspective)

From poverty to prosperity

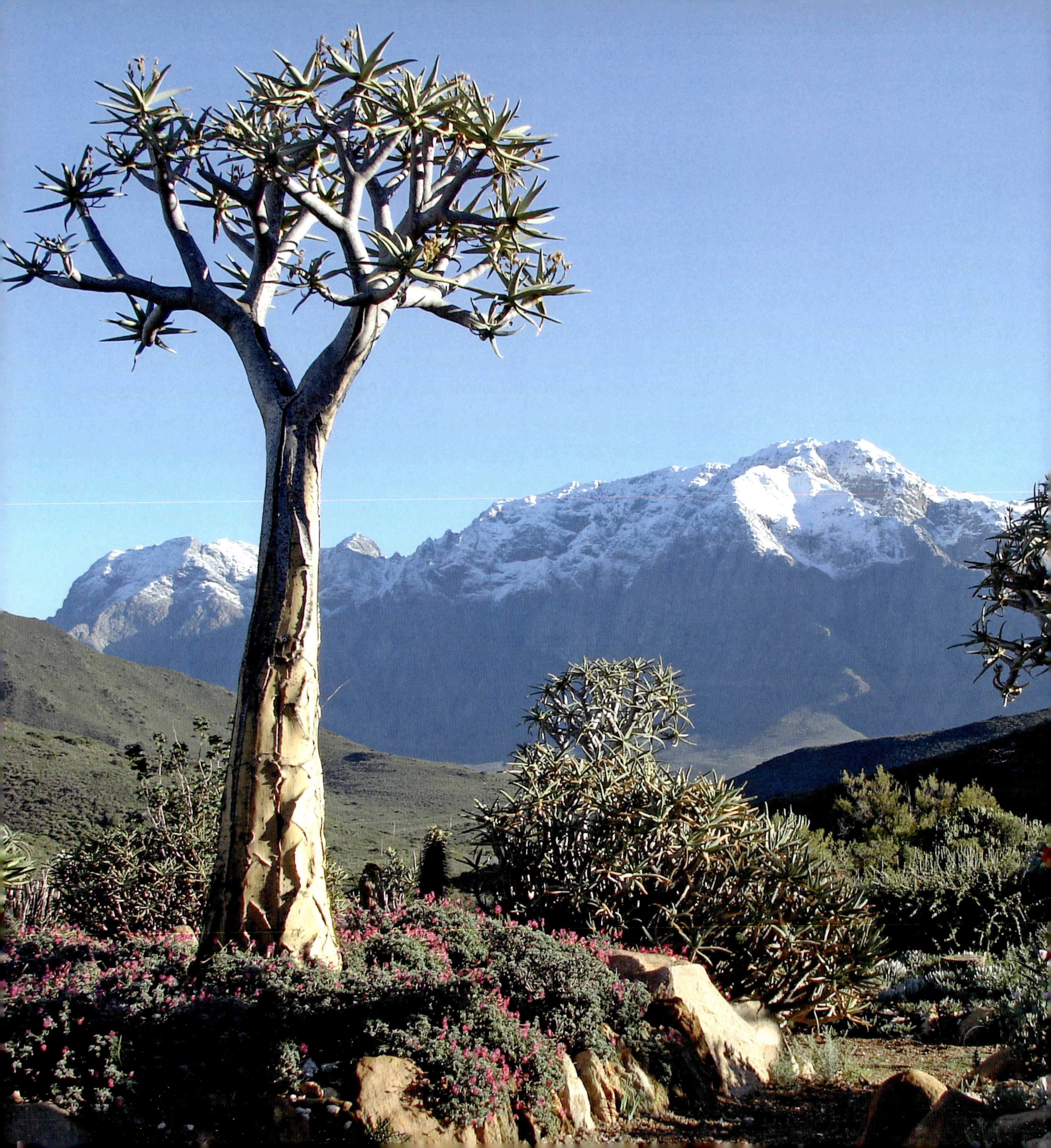

CHAPTER ELEVEN

A network of National Botanical Gardens

Rolling out the dream

Therefore, quite apart from mere geographical considerations, it is obvious that a single garden, however large and well equipped, will not serve the whole of South Africa.

H.H.W. PEARSON
South African Journal of Science, *1910*

The iconic Quiver tree Aloe dichotoma *and other succulents in the Karoo Desert NBG grow below the snow-capped Brandwag peak outside Worcester.*

Lise Wolfaardt

Founding the Network

Harold Pearson's vision of a national network of botanic gardens was not to be realised during his lifetime, nor during that of his successor, Robert Compton. But Pearson had recognised the need for at least one garden in each ecological region of the country – he thought that 10 might suffice. This plan was clearly in competition with that of Dr I.B. Pole Evans, head of the Division of Botany in the Department of Agriculture, based in Pretoria, who had similar ambitions, as discussed in chapter 2. Compton, ever the gentleman, clearly decided against conflict. He did not press for more gardens, aside from the tiny Karoo NBG established in 1921, but the full weight of Pole Evans' influence came to his attention with the establishment of a National Herbarium in Pretoria in 1923.

It was, however, the ecological bent of Brian Rycroft, and his ebullient enthusiasm, that founded the network as we know it today. He was helped by the fact that the programme of botanical reserves initiated by Pole Evans had been changed to focus on pasture research, rather than botanical gardens. Rycroft inherited two gardens – Kirstenbosch and the Karoo NBG. This latter garden was originally a small patch of Succulent Karoo alongside the railway line running from Cape Town to the interior, at Whitehill near Matjiesfontein. This site had proved unsuitable and was abandoned, and in 1945 the Karoo Garden was transferred to its present site at Worcester.

Rycroft travelled the length and breadth of South Africa, meeting with academics, municipal officials, conservation agencies and the general public, most particularly members of the Botanical Society. His infectious enthusiasm soon won over many unlikely supporters, most especially the town clerks of Bloemfontein, Roodepoort, Krugersdorp, Harrismith and Nelspruit. Using the fame of Kirstenbosch as a magic wand, he was able to conjure up images of new botanical gardens in often unpromising urban sites. His targets were not always appropriate. Experience was soon to demonstrate (as in the case of the Drakensberg & Eastern Free State NBG near Harrismith, established in 1967 and abandoned in 1984) that easy access to visitors is a key predictor of an NBG's success. He had ignored the wisdom of Pearson's mentor, Kew Director Sir William Thiselton-Dyer, who wrote in 1880:

> *'The site must be conveniently accessible. A garden, however well managed and stocked with interesting and valuable plants, will be sure to languish if withdrawn in consequence of inconvenience of situation from the eye of the residents.'*

As we will see, Rycroft was not easily discouraged. He succeeded in establishing the Harold Porter NBG in 1959, the Free State NBG in 1967 and the Lowveld NBG in 1969, and the transfer of the Natal Botanic Garden to the National Botanical Gardens in the same year. Shortly before his retirement in 1983, he added the Witwatersrand NBG, later re-named the Walter Sisulu NBG, to his list of successes.

Existing and proposed (Kwelera) National Botanical Gardens across South Africa

A nursery for horticulturists

Kirstenbosch has been the leading institution for indigenous plant horticultural training in South Africa for close on 100 years. The Pearson Memorial Hostel was built in 1923 as a residence for 'lady horticulturists', and both formal and experiential training has been a continuous feature of the Garden's activities.

A major boost to specialised training in indigenous plant gardening came in 1966, with the establishment of the Kirstenbosch Scholarship through a generous donation by Yvonne Eustasie Parfitt, a great enthusiast of the Garden. Many of the country's leading horticulturists are graduates of the Scholarship. For the past 20 years, an internship programme of one to two years, and a six-month in-service training course, have strengthened linkages between the work place and higher education institutions. Over 100 students have participated in Kirstenbosch programmes. Many of these are now permanent employees of SANBI, and four are curators of National Botanical Gardens. The emphasis given to training and transformation is reflected in the composition of the leadership teams across the NBGs.

Over the years SANBI has received numerous awards from Higher Education Institutions for the standard of its in-service training. In 2010 SANBI received a merit award from the South African Society for Co-operative Education (SASCE) in recognition of its standards of excellence.

Since 2000, in order to train garden staff lacking formal qualifications, several skills programmes have been developed on topics such as commercial plant production, botanical garden maintenance, alien vegetation removal and supervisory skills. In total 53 staff have participated in these programmes, some progressing to higher skills levels. The past decade has seen a surge of groundsmen staff at Kirstenbosch registering for higher qualifications in horticulture.

Growing a new generation of horticulturists

The NBG leadership team: ABOVE *in 1979 and* RIGHT *in 2012*

A network of National Botanical Gardens 209

Building gardens with skill and luck

The enduring fund-raising problems at Kirstenbosch had shown that finding funds for developing newly acquired gardens was a slow and frustrating process. Fortunately, the pioneer curators and staff of Rycroft's network were a particularly dedicated lot. With limited means, hard work, and with the blood, sweat and tears that every gardener knows too well, the new NBGs moved forward. Each has its particular history, briefly outlined below.

Good horticulture, with sensitive landscaping, is today reflected in the individual personalities of each of the new gardens. Successive curators have shared their knowledge with one another and with fellow horticulturists across the country. Plant material, too, could be exchanged between gardens and its potential tested under different environmental conditions. Joint field trips in search of promising new introductions built up the living collections of the network as a whole, today numbering over 9 000 species, representing almost 50 per cent of the nation's flora.

During the 1990s, the new South African government was well aware of the need to address unemployment in the country. The Reconstruction and Development Programme was established, followed by the Poverty Relief Programme of the late 1990s, which sought labour-intensive projects that could be used to develop not only basic skills in manual labour, but also life skills, such as in personal financial management and health care.

Ever ready to explore new avenues of support, the then NBI approached its parent agency, the Department of Environmental Affairs and Tourism (DEAT), to access funds for its ambitious capital development programme.

With the novel approach of 'strategic opportunism' introduced by NBI CEO Brian Huntley, an extensive infrastructure development programme was mobilised. Planned by Huntley and implemented by Christopher Willis (who had co-ordinated the highly successful SABONET project prior to being appointed Director of all the NBI gardens in 2000) and other colleagues, it encompassed a new Environmental Education Centre in Pretoria NBG (which became part of the new NBI in 1990), Education and Visitors' centres in Lowveld and Free State NBGs, a Restaurant and Conference Centre at Walter Sisulu NBG, and similar infrastructure in Harold Porter and Karoo Desert NBGs, and in the network's latest acquisition, Hantam NBG. During the first 15 years of the post-1994 government, more funding for infrastructure was raised, and more skills developed and jobs created through projects in the NBGs than in the preceding 80 years.

Succulents flourish in the Karoo, such as these yellow daisies Tripteris oppositifolia *with* Pachypodium namaquanum *and* Euphorbia dregeana *in the background.*

210 KIRSTENBOSCH

Karoo Desert NBG – established 1921

Neville Pillans, the talented young botanist who in 1911 introduced Pearson to the site of the future Kirstenbosch, was also party to identifying the site of the first Karoo NBG. Visiting the arid hinterland of the Cape in 1920, Pillans, Robert Compton, James D. Logan (owner of Rietfontein Estate and known as the 'Laird of Matjiesfontein') and the horticulturist of the South African Railways, Frank Frith, chose an irregular rectangle of land alongside the railway line near Matjiesfontein as the site of Kirstenbosch's first satellite garden. The Garden was named after the railway station, Whitehill. Sir William Hoy, General Manager of the South African Railways, and a passionate friend of Kirstenbosch, authorised free transport by rail from any corner of South Africa for any plant material donated to the Garden. His influence in matters horticultural persisted for many years within the South African Railways system, which once had great pride in its station gardens – now, sadly, a lapsed tradition.

The Whitehill garden had a heroic but short life. Compton's objective was for it to be 'a repository of some of the most remarkable forms of life that the world contains' –

A blaze of bright mesems (mauve Lampranthus hoerlenianus *and red* Drosanthemum speciosum*) surrounds the thatched Kokerboom restaurant at the Karoo Desert National Botanical Garden.*

A network of National Botanical Gardens   211

The living collections of desert succulents in the Karoo Desert NBG are a major scientific and conservation asset.

Brianhuntleya intrusa, an Endangered succulent endemic to the Little Karoo, is propagated for sale by the Garden's nursery.

Yellow daisies Hirpicium integrifolium *and pink* Ruschia caroli *bring colour to the harsh Karoo landscape.*

a reflection on the fascinating flora of the arid Karoo. Funds were extremely limited, as was rainfall – an average of less than 250 millimetres per annum. The first Curator was the local Station Master, Joseph Archer, appointed in 1925, who served the Garden until 1939. His enthusiasm, and the interest of the botanists at Kirstenbosch, resulted in a paper being published on the flora of the Whitehill district, which included the description of an astonishing four new genera and 52 new species. The Garden struggled on without a Curator once Joseph Archer had left, until the decision was taken in 1945 to move it to a more accessible site – in Worcester. The Worcester Municipality, and a donor, Charles Heatlie, made land available at the foot of the Brandwacht Mountain on the margins of the pretty little farming village (as it was then). The move took place in 1945/46, with Swiss horticulturist Jacques Thudichum as Curator from 1945 to 1958.

The official opening of the new Karoo National Botanic Garden, attended by 600 people, was presided over by the Governor General, the Right Honourable Gideon Brand van Zyl; he had been a member of the Board of Trustees of the NBG for several years, and had, in fact, drafted the original constitution of the Botanical Society back in 1913. The audience included the Prime Minister, Field Marshal Jan Christiaan Smuts.

Jacques Thudichum laid the foundation to the Garden; he was followed by a succession of succulent enthusiasts including taxonomist/horticulturist Bruce Bayer, who built up the Garden's scientific collections and developed the stunning spring displays of iridescent vygies (or mesems) for which the site is world renowned.

The 154-hectare Garden lies on undulating hills and plains of Malmesbury Shale, 300–520 metres above sea level. It occupies vegetation of both the Fynbos and the Succulent Karoo biomes – Breede Shale Fynbos, Breede Shale Renosterveld and Robertson Karoo. Eleven hectares of the Garden are developed as displays of the huge diversity of the Succulent Karoo flora, with 8 kilometres of footpaths leading the visitor through the Garden's show beds and up into the surrounding hills. Despite the Garden's small size and the harsh climate (250 millimetres of annual rainfall, with temperatures ranging from -2 to 44°C), 370 indigenous species have been recorded, 83 of these geophytes, and 74 true succulents. The single genus *Crassula* has 20 species; *Pelargonium* is represented by 13 species; there are an astonishing 27 different members of the Mesembryanthemaceae, plus seven xerophytic ferns; all are native to the Garden – in a desert ecosystem. To this array of diversity, some 4 000 other species (65 per cent of which are succulents) have been added by the Garden's dedicated horticulturists – not too difficult a task, given that, within a radius of 100 kilometres of the Garden, there are 65 different vegetation types.

Kokerbooms tower above the Aloe pearsonii *(named after first Kirstenbosch Director, Harold Pearson) in the foreground.*

A network of National Botanical Gardens

Awash with spring colour: the orange daisy Gazania rigida *shares the landscape with the white-flowered* Dimorphotheca pluvialis.

214   KIRSTENBOSCH

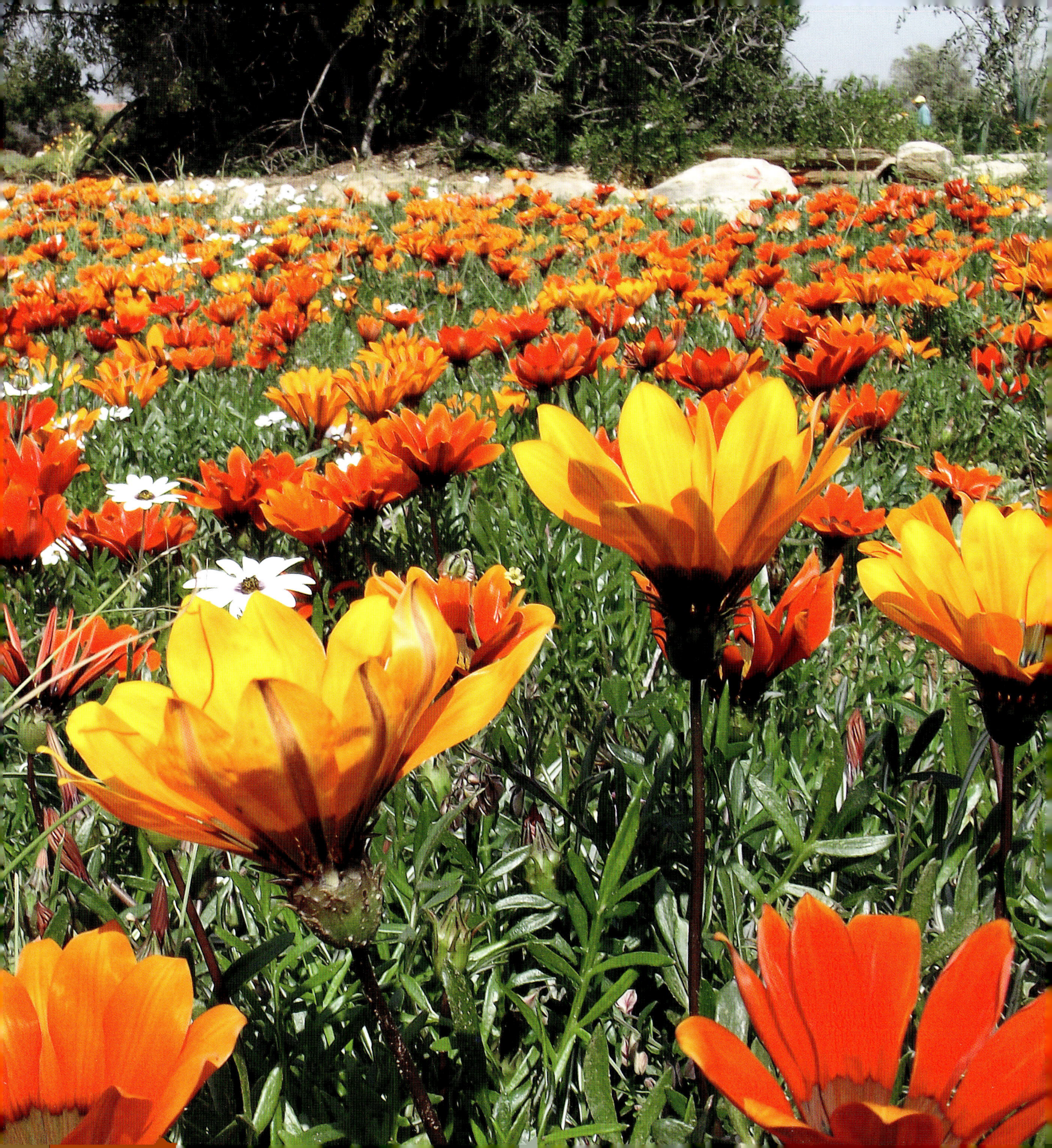

Pretoria NBG – established 1958

In 1923, when Jan Smuts declared that the new National Herbarium in Pretoria could become the 'Kew of South Africa', he triggered an institutional struggle of serious proportions, and one that took many decades to heal.

The National Herbarium of the 1920s was housed in an ageing building 'Vredehuis', below the Union Buildings, on a site offering limited potential as a botanical garden. Dr Illtyd Buller Pole Evans, chairman of the Botanical Survey Advisory Committee and leading light of botany in South Africa at the time, realised that a southern hemisphere 'Kew' would need a large, well-developed botanical garden, along with a herbarium and research facilities. Occupied as he was with the development of many other projects, Pole Evans did not pursue the search for such a site in Pretoria. This task fell to one of his early assistants, Robert Alan Dyer (1900–1987), a taxonomist of wide interests and great productivity. In 1946 Dyer became Director of the Division of Botany (the Botanical Research Institute from 1961) and was the driving force behind the establishment of the Pretoria NBG. He negotiated the transfer of 60 hectares of land donated by the University of Pretoria to the Department of Agriculture in 1945. Declared unsuitable for inclusion in the University's Experimental Farm, the sandy, acidic soils nevertheless showed promise for development as a botanical garden. Over the following decades, other adjoining land was purchased for inclusion in the Pretoria NBG, which today totals 75 hectares in extent.

Located some 10 kilometres from the then offices of the National Herbarium at Vredehuis, and situated beyond the outskirts of the city, the Pretoria National Botanical Garden was officially opened in 1958. Despite its name, its primary purpose in the early days was to provide researchers – taxonomists working at the National Herbarium – with living material, and the public was not permitted daily access until 1984.

The Garden runs along a wooded ridge of quartzite, the eastern extension of the Magaliesberg ridges that separate the cooler southern from the warmer northern suburbs of the capital city. The natural vegetation of the Garden, Marikana

An Ndebele village adds interest to the Medicinal Plants garden in Pretoria National Botanical Garden.

An avenue of the beautiful Tree Wisteria Bolusanthus speciosus *was one of the first plantings in the Pretoria NBG.*

Thornveld, has been almost completely transformed through past farming practices and the development of the Garden landscape, but the quartzite ridge retains most of its bushveld tree and shrub species that are typical of much of the northern provinces of South Africa. Many of the trees and shrubs, such as Wild Seringa *Burkea africana*, Peeling-bark Ochna *Ochna pulchra* and Stamvrug *Englerophytum magalismontanum* extend in their range from Pretoria to the margins of the Congo Basin some 3 000 kilometres to the north.

Because the Garden's primary objective was for research, large glasshouses were built in the 1970s and were devoted to masses of potted plants used in experiments or kept as living collections – vast arrays of succulents, orchids and Madagascan plants that are still preserved with great dedication in the now rather decrepit glasshouses, to which the general public has never had access.

It was not until the 1990s that the Garden developed a friendly face. Building on efforts made in the 1980s, theme gardens were developed, representing the different biomes of South Africa, with a selection of fynbos species hardy enough to survive the cold, dry winters and hot, wet summers of the area. Bushveld trees and aloes have been successfully established along the warm north-facing slopes of the quartzite ridge, while an enchanting forest with a discreetly engineered waterfall provides a dense, shady retreat for visitors from the summer sun.

An artificial waterfall creates a cool, moist habitat on the south-facing slope of the Pretoria NBG's quartzite ridge.

Despite the challenges – among them, of securing an abundant and inexpensive water supply – the Garden has become a verdant oasis in the midst of a now highly urbanised landscape. Theme gardens, visitors' facilities, a large and active environmental education programme, concerts, conferences and masses of happy visitors have transformed it from a staid research institution to a great public asset.

A brand new Biodiversity Centre was built in 2004 alongside the National Herbarium, to house the new Head Office of what soon became the South African National Biodiversity Institute. From its humble beginnings as a herbarium collection, botany, and indeed biodiversity, has now found a home in the delightful surroundings of the Pretoria National Botanical Garden.

Harold Porter NBG – established 1959

In 1940, Harold Porter, an architect and town developer with a keen interest in plants (having been Chairman of the Transvaal Horticultural Society in 1925), obtained rights to a large stretch of land at the mouth of a pair of narrow valleys running down to the sea from the rugged peaks of the Kogelberg. In 1950 he started work on what is today a little gem of a garden, the Harold Porter National Botanical Garden. He was not to live to see his dream fully realised, but, after his death in 1958, his garden was donated to the National Botanic Gardens, and incorporated into the nascent national network in 1959.

The Garden today covers 200 hectares and includes four distinct vegetation types – Hangklip Sand Fynbos, Kogelberg Sandstone Fynbos, Overberg Dune Strandveld and Southern Afrotemperate Forest. Fynbos covers the steep slopes of the Kogelberg, with pockets of Afrotemperate Forest in

The red-flowering Kogelberg endemic Brunia stokoei *blooms brightly against the backdrop of the Leopardskloof.*

218 KIRSTENBOSCH

the deeply incised valleys of the Disa and Leopard's Kloof streams. The streams flow through a dense wetland of Palmiet *Prionium serratum* before reaching the sea across a narrow band of Dune Strandveld.

The Garden has seen a succession of 12 curators during the half century of its development. Somewhat miraculously, none of them has tried to force changes upon the design of the 10 hectares of cultivated garden. Fires and floods – some of devastating proportions – have kept curators and the team of gardeners focused on maintaining a simple but effective formula for success. A well-designed circular path, with a second path radiating deeper into the valleys and over a mountain spur, provides visitors with a most delightful walk through theme gardens of coastal and mountain fynbos, dune, forest, wetland and, most recently established, limestone-fynbos ecosystems. Easily accessible walkways lead visitors to the deep dark amber pools of the Disa stream, edged by large specimens of Cape Holly *Ilex mitis*, Cape

The epicentre of floral richness

In January 1773, during one of his wanderings in search of new plants for Kew, the indefatigable Francis Masson ascended the steep rocky pass over the Hottentots Holland Mountains and descended into the valley of the Palmiet River. Of this area, the Scottish gardener had the following to say:

'These mountains abound with a great number of curious plants, and are, I believe, the richest mountains in Africa for a botanist.'

Masson was entering the Kogelberg, which botanists – more than two centuries later – still regard as the epicentre of the floral richness of the Cape Floral Kingdom. Kogelberg was included as part of the Cape Floristic Region World Heritage Site inscribed by UNESCO in 2002 (along with Kirstenbosch and the Cape Peninsula). It was here, on the coast below the high mountains of the Kogelberg, that a small private botanical garden was established in 1950, and aptly named 'Shangri-La' – a place of peace and harmony. This stretch of land was to become the future Harold Porter National Botanical Garden.

The southernmost outliers of Afrotemperate Forest occur along the streams and deep dark waters of the Disa and Leopard's kloofs.

A network of National Botanical Gardens   219

The Critically Endangered Match Stick Pagoda Mimetes hottentoticus *is endemic to the Kogelberg.*

Beech *Rapanea melanophloeos*, Hard Pear *Olinia ventosa*, Rooiels *Cunonia capensis* and Keurboom *Virgilia capensis*. At the head of the valley a high waterfall cascades down into a small pool, above which the Red Disa *Disa uniflora* may be seen flowering during midsummer.

The Kogelberg flora has 1 882 plant species, many of these endemics, and including representatives of the Cape endemic families Penaeaceae and Bruniaceae, which can be seen in the Garden. The mountainsides are rich in restios, ericas and proteas, with fine specimens of the country's national flower, the King Protea *Protea cynaroides*, being a special feature.

South Africa's national flower, the King Protea Protea cynaroides, *grows abundantly in the Harold Porter National Botanical Garden.*

Free State NBG – established 1967

When Kirstenbosch celebrated its Golden Jubilee in 1963, the then Director, Brian Rycroft, invited 50 eminent botanists from around the world to join the party. Following the Jubilee events in Cape Town, the group travelled widely throughout South Africa, and included a stop in Bloemfontein, affectionately known as the 'City of Roses'. Rycroft and his illustrious guests saw the need for a garden representing the vegetation and flora of the central highveld, and his energy was rewarded with the establishment of the Free State National Botanical Garden in 1967 – the first NBG beyond the borders of the then Cape Province.

Lying at an altitude of 1 400 metres above sea level, the Garden is 67 hectares in area. The natural vegetation is adapted to the rather harsh conditions of the highveld plateau. Extreme temperatures recorded in Bloemfontein range from -10°C in July and August to over 39°C in February, with frequent frosts through winter. Rainfall is highly seasonal, from October to March, with an average of 550 millimetres per annum.

Most of the Garden comprises lightly wooded rolling slopes and grasslands, with wooded dolerite hills in the northern section. A shallow valley, dammed by British troops to water their horses during the South African War of 1899 to 1902, provides a semi-permanent water feature.

The two vegetation types found in the Garden, Winburg Grassy Shrubland and Bloemfontein Karroid Shrubland, represent the two biomes, Grassland and Nama-Karoo, that dominate the region. The winter landscapes are dry and brown, broken here and there by radiant aloes and cotyledons and the evergreen trees and shrubs of the dolerite hills. The first rains of spring turn the Garden into a green oasis, decorated by many grassland geophytes – boophone, brunsvigia, nerine, haemanthus, ammocharis and hypoxis; and by the new growth of trees and shrubs – False Olive *Buddleja saligna*, African Olive *Olea europaea*, Blue Guarri *Euclea crispa*, Cross-berry *Grewia occidentalis*, Mountain Cabbage-tree *Cussonia paniculata*, and Karee *Searsia lancea* (formerly *Rhus lancea*), among others – that provide shade in the hot summers.

The Free State NBG lies in the rolling hills of the central highveld. Hardy evergreen trees survive the cold, frosty winters, and summer rains fill an old dam built during the South African War (1899–1902).

A traditional 'Hartebeeshuis' has been constructed in the Useful Plants garden.

A network of National Botanical Gardens

Given the harsh climate and limited supply of water, only 7 hectares of the garden have been landscaped, and are dedicated to the flora of the Free State. Following the tradition of theme gardens, the Garden has a Medicinal Plants garden, complete with two traditional houses typical of the area – a 'Hartebeeshuis' from the Boer culture, and a Sotho hut. The water-wise garden is highly appropriate to this drought-prevalent area. A 'Garden of Hope' has been developed with the support of the Botanical Society. Created as a symbol of hope for people suffering from HIV and AIDS, this garden offers a sensory experience in which visitors can relax, reflect and heal in beautiful and peaceful natural surroundings.

Once perceived as the 'Cinderella' of the network of NBGs, the Free State NBG has benefited from a surge of financial support from the government's Expanded Public Works Programme. Since 2002, an Environmental Education Centre, Visitors' Centre, Plant Sales Nursery, restaurant and parking area have been constructed, following a sensitive design provided by local architect and garden enthusiast Jan Ras. The works programme continues with many labour-intensive projects, including alien plant removal, laying of pathways, upgrading of irrigation and general maintenance.

Lowveld NBG – established 1969

In his search for sites for new NBGs, Brian Rycroft clearly had an eye for the dramatic as well as for the botanically fascinating. Few of his discoveries could have been more rewarding than the visually spectacular gorges cut deep into the landscape just outside the city of Nelspruit, capital of Mpumalanga province. Here the Crocodile and Nels rivers have carved sheer ravines that drop precipitously below the rolling hills of this beautiful bushveld scenery. The area's subtropical climate, with 750 millimetres of rainfall per year, and no frosts, plus abundant water from the Crocodile River and deep, rich soils, provides ideal gardening conditions. The dominant vegetation of the Garden, Pretoriuskop Sour Bushveld, takes its name from the popular tourist rest camp just inside the Kruger National Park, to which the Garden serves as an appropriate gateway.

The land for the Lowveld NBG was donated by H.L. Hall & Sons, owners of a vast agricultural estate in the area, while additional land was contributed by the Nelspruit Municipality. The 165-hectare Garden has 25 hectares of developed plantings, where emphasis has been given to collections of trees and shrubs of South Africa. The natural vegetation includes 470 species, with a further 2 020 species in the Garden's living collections. The use of shade-tolerant Durban Grass

An abundant diversity of trees grows in the African Rain Forest.

The Visitors' Centre at the Lowveld National Botanical Garden is set in the woodland above the deep Crocodile River gorge.

The Crocodile River cascades spectacularly through the Lowveld National Botanical Garden.

Dactyloctenium australe results in lush evergreen lawns below shady bushveld trees: a diverse savanna parkland sweeping across the Garden's gently rolling shoulders before dropping into the deep gorges of the Crocodile and Nels rivers.

Johan Kluge (1947–1998), long-time Curator of the Garden (from 1980 to 1998), had a special passion for trees, and established two forests – the African Rain Forest and the South African Forest – that present an abundant diversity of the woody plants from these highly threatened habitats of West, Central and South Africa. Funding provided by Sappi, a pulp and paper company, saw the realisation of one of Kluge's dreams – a raised walkway that takes visitors through the forest canopy. This innovation adds another dimension to the visitor's experience – a closer look at some of the Garden's 243 bird and 96 butterfly species.

Many discreetly placed viewing sites allow access to the lip of the gorges, and to views of the rivers in full spate as they tumble over high waterfalls into the deeply potholed river below. A suspension bridge links the two sides of the Garden, giving access from the Visitors' Centre, restaurant and parking to the Environmental Education Centre and the main Garden with its outstanding collection of cycads, clivias and fig trees.

While trees are the main focus of the Garden's collections, the conservation of cycad gene banks has been a key activity, with large and well-protected orchards of 53 cycad species planted out to provide seeds for mass propagation and distribution to other gardens and conservation agencies.

Easily accessible from the main national roads linking the interior of the country with Mozambique and the Kruger National Park, the Lowveld NBG is a tourist destination that should not be missed.

A network of National Botanical Gardens

KwaZulu-Natal NBG – established 1969

In the rush to expand the national network of Botanic Gardens, so energetically pursued in the 1960s, it is perhaps regrettable that an old, rather moribund colonial garden was drawn into the fold. Natal Botanic Garden in Pietermaritzburg had been established in 1874. In common with the many similar gardens being developed in the distant reaches of the British Empire, its primary objective was to serve as an acclimatisation garden for the introduction of timber and other economically important species suitable for use in the young colony of Natal. It served this purpose well: many thousands of trees, mainly wattles and eucalypts, were supplied by the Garden through the late 19th century to government buildings, schools, hospitals, cemeteries and private individuals. Funds were always lacking and, despite the rather elegant Victorian design and displays of hydrangeas, rhododendrons, azaleas, bougainvilleas, poinsettias and other horticultural favourites of the time, by 1900, in the view of some, 'the Gardens have in the main become pleasure gardens and not botanic gardens, a place for lounging rather than learning'.

But the Garden had some fine curators, including W.E. Marriot, who in 1908 planted the 200-metre double

The double avenue of London Plane trees, planted in 1908, provides a popular venue for parties and family outings at the KwaZulu-Natal NBG.

Winter concerts are a major drawcard for visitors to the KwaZulu-Natal National Botanical Garden.

avenue of the London Plane tree *Platanus x acerifolia*, still to this day the Garden's most charming asset. In 1907 the *Natalian* had lauded Marriot's work:

> 'The great improvements that have taken place in the last two years are rapidly raising the Garden to a position second to none in South Africa ... a veritable lung of the city, soothing to the eye and brain, and appealing to the sense of all that is beautiful and good and noble.'

Despite serious ups and downs, in 1930, the eminent botanist and scientific advisor to the British government, Sir Frederick Keeble (1870–1952), considered it 'An ideal Botanical Garden – the Pride of Natal and of Kew'.

But by 1969 the Garden was in dire financial trouble, and the Pietermaritzburg Botanic Society readily transferred it to the expanding umbrella of the National Botanic Gardens. Peter Law, the first Curator appointed by NBG to the Garden, immediately set about eradicating invasive species that had infested much of the natural area of the Garden. By his retirement in 1979, most of the worst invasive trees had been removed, with many trees and shrubs indigenous to the region taking their place. Poor shallow soils and cold winters have made the re-establishment of the original vegetation – indigenous Southern Mistbelt Forest and Ngongoni Grassveld – a slow and difficult task. Alien invasive species find the habitat ideal, and the fight to control their infestations is endless, sapping the energies and resources of successive curators and staff.

Transforming the original colonial garden into an indigenous garden has also been problematic, its 'old bones' proving difficult to disguise. Many of the rigid, geometrically patterned plantings of trees remain, including some fine specimens of Camphor *Cinnamomum camphora*, Swamp Cypress *Taxodium distichum* and Moreton Bay Fig *Ficus macrocarpa* – indications of the Garden's Victorian heritage. Despite the best efforts of its curators, the problems resulting from a long and changing history continue: the Garden lacks the 'spirit of place' that each of the other NBGs offers.

As SANBI's mandate requires increasing attention to biodiversity rather than to plants in isolation, the Garden offers some compensation for its other shortcomings. It is a haven for birds (182 species), butterflies (126 species), dragonflies (26 species) and damselflies (15 species). The butterfly fauna alone represents 29 per cent of the province's butterfly diversity, in an area totalling less than 0.0005 per cent of the province. Bearing these good points in mind, former Curator (1979–2009) Brian Tarr wrote in positive vein in 1979, 'Natal Botanic Garden – an old garden with a bright future'.

A network of National Botanical Gardens

CLOCKWISE FROM TOP LEFT *Floral magic from the NBGs:* Drosera cistiflora; Leucospermum cordifolium; Cyanotis speciosa *and* Orothamnus zeyheri

CLOCKWISE FROM TOP LEFT *Insect beauties from the NBGs: the Brown-veined White, Mother-of-Pearl and Green-veined Charaxes butterflies and the Darting Cruiser dragonfly*

A network of National Botanical Gardens   **227**

Walter Sisulu NBG – established 1982

One of Rycroft's final tasks before retiring in 1983 was to establish a new garden in the country's industrial and financial capital, Johannesburg. The existing municipal garden at Emmarentia was not available for transfer to the NBG – a blessing in disguise. Like the Natal Botanic Garden, it lies on a difficult site and has all the problems of an old garden in a new society. What Rycroft found was a perfect site, long used by picnickers of the suburbs and towns radiating westwards from the centre of Johannesburg. Located along the Crocodile River, which separates the municipalities of Roodepoort and Krugersdorp, it occupies the transition from the cold, treeless plateau of the Witwatersrand to the warmer slopes and valleys of the

A pair of Verreaux's Eagles has nested on the cliffs next to the waterfall; they are jealously protected as icons of the Garden.

The Witpoortjie waterfall is the centre point of the Walter Sisulu National Botanical Garden.

228　KIRSTENBOSCH

bushveld. Receiving 760 millimetres of rainfall per year, with warm summers and cold winters, the Garden is vulnerable to the occasional hard frost. Remarkably, the major portion of this 276-hectare tract of land, in the middle of the richest gold fields in the world, was in an almost pristine state when it was made available to the NBG in 1982.

Known, briefly, as the Transvaal NBG, and then the Witwatersrand NBG, the Garden was renamed in 2004 in honour of the late ANC President Walter Sisulu (1912–2003), who, with Nelson Mandela, led the struggle for a democratic South Africa.

The centrepiece of the Garden, hidden from the entrance and unfolding as one advances along the main circular walkway, is the spectacular Witpoortjie Waterfall, a 70-metre-high natural feature tumbling over a cliff that reveals the area's rich geological history. Nesting high on the cliff face adjacent to the waterfall is a breeding pair of majestic Verreaux's Eagles – an image of which species is the Garden's much-treasured icon.

The aloe collection at the Walter Sisulu NBG is one of the best in the country.

A small central area of the Garden had been used for agriculture during the first half of the 20th century. The task of removing the many retaining walls, cottages and farm buildings, and that of establishing the hard landscaping and basic infrastructure of the Garden, was completed in less than five years by the Garden's first Curator, Peter Chaplin. The framework put in place has matured into a superb asset for the region, so that it is repeatedly voted the best ecotourist destination in Gauteng.

The first wave of infrastructural development was achieved through fund-raising among the many large industrial and mining houses of the region. A wetland was established, with a bird-watchers' hide, using funding from Sasol; an Environmental Education Centre was funded by Nestlé; and the main circular walkway was donated by Anglovaal. The second wave came with the help of the Expanded Public Works Programme – a new restaurant, Visitors' Centre and Conference Centre were completed in 2008.

The main vegetation of the Garden is Gold Reef Mountain Bushveld – a mosaic of grassland and savanna, with a dense woodland of White Stinkwood *Celtis africana* along the course of the Crocodile River. The slopes of the northeast-facing escarpment that forms the backbone of the Garden have mixed savanna woodland, with a great diversity of bushveld trees, grasses and forbs and, remarkably, the last surviving population of Mountain Reedbuck in Gauteng. Some 458 plant species grow naturally in the Garden, with another 852 species in the living collections.

The soils and climate have proved suitable for aloes and other succulents, and the winter display of reds, golds, mustards and yellows of the dozens of species of aloe in the Garden is among the best in the NBG network. Birds are prolific, with no fewer than 243 species having been recorded here.

A unique and appropriate feature in the Garden is the Geological garden. Built with the co-operation of the Geological Society of South Africa, and with the support of several mining houses, the garden displays massive specimens of the geological formations that underpin the country's history, and its economy, as one of the richest mineral producers globally.

A network of National Botanical Gardens

Hantam NBG – established 2008

When Masson and Thunberg trekked through the remote and rough hills and plateaus of the Bokkeveld in 1774, they had no expectation that the fascinating plants that they were collecting would become the subject of international interest two centuries later, nor that the region would become known as the 'Bulb Capital of the World'.

The whole of the west coast of South Africa has long been known for its spectacular spring wild-flower displays. Flower tours of Namaqualand, the Knersvlakte and the Bokkeveld Plateau have been popular for many decades. The cold Bokkeveld Plateau, lying at between 750 and 1 400 metres above sea level, is home to 1 350 species of plant, including 83 narrow endemics, found here and nowhere else. What makes the Bokkeveld Plateau especially interesting is the massive diversity of bulbous plants (geophytes) to be found in this small area. Some 309 species, in the families Amaryllidaceae, Iridaceae and Hyacinthaceae, have thus far been recorded – more than in the combined Mediterranean-climate regions of California, Chile and Australia. The stunning diversity of its flowers and nature's extravagant displays have made the Bokkeveld Plateau a popular destination among plant scientists.

A local farm, Glenlyon, was owned by the MacGregors, a family of Scottish settlers who had purchased it in 1883. The farm was made famous a century later by Neil MacGregor (1936–2010) who developed highly original conservation farming practices that ensured an ever-improving display of the farm's spectacular wild-flower diversity. MacGregor was well known for his infectious enthusiasm and gracious charm, over the years taking thousands of visitors around the flower spectacles on Glenlyon in an ancient Bedford bus. Such was the value of the knowledge shared by MacGregor that the NBI, from 1999 to 2004, included Glenlyon in a national project that studied best practice in farm management in terms of biodiversity conservation. When the NBI was searching for a site on which a new NBG could safeguard the unrivalled floral richness in the Northern Cape, Glenlyon was clearly the top candidate and the ultimate site of choice.

The spring flower spectacle at the Hantam National Botanical Garden draws visitors from around the globe.

Spring flowers bring the Hantam National Botanical Garden to life.

230 KIRSTENBOSCH

Partnerships in Nature

Research by Kirstenbosch botanists John Manning, Kim Steiner, Peter Goldblatt and many associates during the 1980s had spread the fame of the Bokkeveld far and wide. In the early 1990s the BBC produced the TV documentary series *Private Lives of Plants*, enchantingly narrated by Sir David Attenborough at his most inspired. The subject of the documentary was the intriguing co-evolution of plants and animals. At the farm Glenlyon, just outside the town of Nieuwoudtville, they filmed the pollination of the long-tubed corollas of *Lapeirousia* and *Babiana* by long-tongued flies.

Another fascinating relationship discovered was that of oil-collecting bees, attracted to *Diascia* and *Hemimerris* flowers, which have evolved tiny jars of oil in specially elongated spurs of the corolla. The bees insert their hairy front legs into the oil-filled spurs of the flowers, simultaneously collecting oil and pollinating the flower.

Yet another fascinating discovery was that of the association of monkey beetles and the brilliantly coloured irid *Romulea*. Different species of monkey beetle are specific to different species of very similar-looking *Romuleas* growing on different soils, such as *Romulea sabulosa* on tillite soils and *R. monadelpha* on dolerite soils. It appears that the ecological separation of the different species of monkey beetle associated, in this instance, with the two species of *Romulea*, is a consequence of the choice of soil type by the beetle larvae, and not due to any particular feature of the plants with which they are so often found associated.

A long-tongued fly pollinates the narrow tube of Babiana framesii.

Two similar species of Romulea *are pollinated by different species of monkey beetle, which select their hosts according to soil conditions – not flower pattern.* R. sabulosa (LEFT) *occurs on tillite clays, and* R. monadelpha (RIGHT) *on dolerite soils.*

A network of National Botanical Gardens

Every spring, a seemingly endless carpet of flowers reaches to the horizon in the Hantam National Botanical Garden, a world-famous floral festival. Seen here are Hesperantha pauciflora *(pink),* Oxalis obtusa *(pale yellow) and* Felicia australis *(blue).*

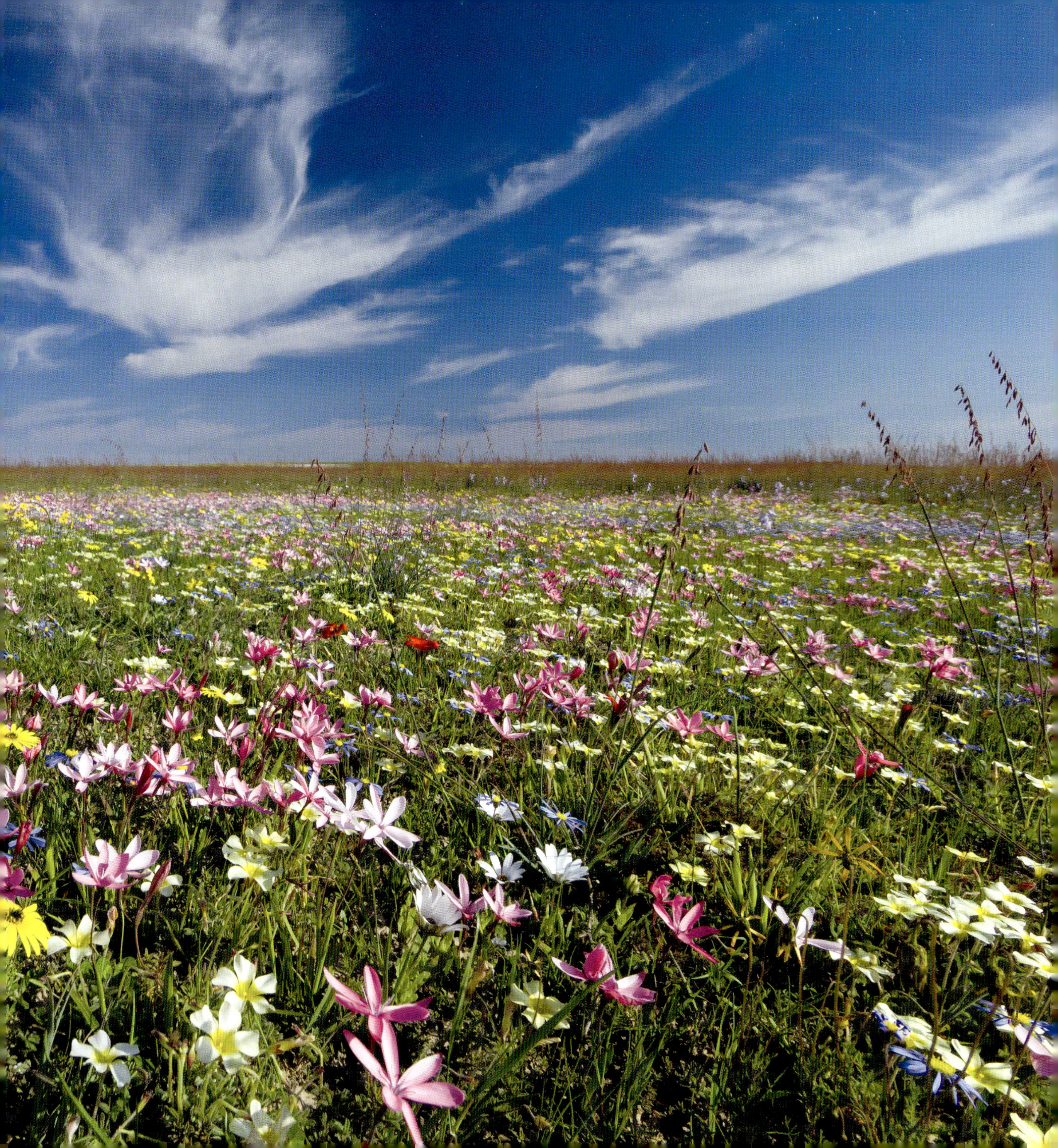

In autumn, the dry and desolate landscape is transformed into a wonderland of pinks with the mass flowering of Brunsvigia bosmaniae.

When funds were sought to purchase the farm, the Botanical Society, WWF South Africa and individual local donors could not meet the R10 million needed. Jorgen Thompson of Conservation International, who had earlier facilitated a gift of R12 million from the Rufford Maurice Laing Foundation to build the Centre for Biodiversity Conservation at Kirstenbosch, once more came to the rescue by paving the way for a R5-million gift from John and Kirsten Swift, California-based supporters of CI's Global Conservation Fund. Further funding from the Leslie Hill Succulent Karoo Trust, through WWF-SA, and from the Department of Environmental Affairs and Tourism, secured the deal. On 28 August 2007 a launch celebration was held at Glenlyon, which delighted all sponsors with a splendid display of wild flowers. In December 2008 it was proclaimed the ninth National Botanical Garden of South Africa.

The area's vegetation comprises successive bands of Nieuwoudtville Shale Renosterveld, Nieuwoudtville-Roggeveld Dolerite Renosterveld and Hantam Karoo. The vegetation pattern follows gradients of geological substrate and of decreasing altitude and rainfall eastwards from the edge of the escarpment, above which the Bokkeveld Plateau is perched. While the escarpment receives as much as 800 millimetres of rain per annum, the Garden receives, on average, only 350 millimetres.

Most of the Hantam NBG lies on the Ecca shales of the Hantam Karoo, but the richest diversity of plants and animals is found on the Dwyka tillite sediments, lain down during the Karoo geological period, some 300 million years ago. The tillites form hard clayey soils, poorly drained and slightly acidic, that become waterlogged during the wet winters. Renosterveld is found on these soils, which have high agricultural value for rooibos tea production and of which few areas remain untransformed. It is on these soils that the immense richness of the area's bulbs thrives. Studies undertaken during the Conservation Farming Project revealed that the density of bulbs and bulblets can be as high as 25 000 per square metre, although densities of a few hundred plants, of up to 50 species per square metre, are more common. Such high densities are needed in ecosystems where molerats and porcupines may eat several hundred bulbs a night.

Gardens of the future

By most standards, one might consider the South African network of NBGs as adequate to provide for both the representation of the country's extraordinary botanical diversity and for the conservation of threatened plants. But SANBI's Gardens Expansion Strategy, developed and approved in 2010, aims to have a National Botanical Garden established in every province of South Africa by 2020. New gardens are planned for the Eastern Cape, Limpopo and North West provinces.

Two regions, the botanically rich but threatened Eastern Cape coast, which falls within the Maputaland/Pondoland biodiversity hot spot, and the mountains of the Soutspansberg in Limpopo province, where outliers of Afrotemperate Montane Forests occur within savanna and woodland complexes, are particularly deserving of inclusion in the network.

Harold Pearson's vision of at least 10 NBGs, which he outlined in 1910, will very probably be met by the centenary of Kirstenbosch's establishment. Remarkably, his listing of priorities for new gardens included seven of the current nine NBGs, plus the Eastern Cape, despite his concern that:

'The compilation of a complete list of these will only be possible when we have gained experience which is not available today.'

Postscript – a new NBG in the Eastern Cape

Pearson's vision is, in fact, due to be realised within the Kirstenbosch Centenary. Through a formal co-management agreement signed on 30 March 2012 by SANBI CEO Tanya Abrahamse, the Eastern Cape Parks and Tourism Agency (ECPTA) and the Eastern Cape Department of Economic Development, Environmental Affairs and Tourism, the country's 10th NBG will be established by 2013. The site is located on the coast 20 kilometres north of East London and includes marine frontage between the Gonubie and Kwelera rivers, as well as pristine coastal dune forest and grasslands. Appropriately, the Kwelera National Botanical Garden derives its name from the Khoikhoi word meaning 'many aloes'.

Pristine coastal dune forest forms part of the proposed Kwelera National Botanical Garden, due for establishment by 2013.

Bibliography

Arnold, M. (Editor). 2001. *South African Botanical Art: Peeling back the Petals*. Fernwood Press, Cape Town.

Ashwell, A., Sandwith, T., Barnett, M., Parker, A. & Wisani, F. 2006. 'Fynbos Fynmense – people making biodiversity work'. *SANBI Biodiversity Series 4*. South African National Biodiversity Institute, Pretoria.

Bradlow, F.R. 1994. *Francis Masson's account of three journeys at the Cape of Good Hope 1772–1775*. Tablecloth Press, Cape Town.

Burchell, W.J. 1822 & 1824. *Travels in the Interior of South Africa, Volumes 1 & 2*. Longman, Hurst, Rees, Orme & Brown, London.

Cadman, M., Petersen, C., Driver, A., Sekhran, N., Maze, K. & Munzhedzi, S. 2010. *Biodiversity for Development – South Africa's landscape approach to conserving biodiversity and promoting ecosystem resilience*. South African National Biodiversity Institute, Pretoria.

Compton, R.H. 1965. *Kirstenbosch – Garden for a Nation*. Tafelberg Uitgewers, Cape Town.

Compton, R.H. 'A botanist's distant reminiscences'. *Veld & Flora*, December 1976.

Cowling, R. & Richardson, D. 1995. *Fynbos – South Africa's Unique Floral Kingdom*. Fernwood Press, Cape Town.

Frazer, M. & Frazer, L. 2011. *The Smallest Kingdom: Plants and Plant Collectors at the Cape of Good Hope*. Kew Publishing, Richmond.

Goldblatt, P. & Manning, J. 2000. 'Cape Plants. A conspectus of the Cape Flora of South Africa'. *Strelitzia 9*. South African National Biodiversity Institute, Pretoria.

Gunn, M. & Codd, L.E.W. 1981. *Botanical Exploration of Southern Africa: An Illustrated History of Early Botanical Literature on the Cape Flora*. Balkema, Cape Town.

Fagan, G. 1973. 'Kirstenbosch Development Study'. Report to the Board of Trustees, National Botanic Gardens of South Africa, Cape Town.

Hoffman, T. & Ashwell, A. 2001. *Nature Divided: Land Degradation in South Africa*. UCT Press, Cape Town.

Honig, M. 2000. 'Making your garden come alive! Environmental Interpretation in Botanical Gardens'. *Southern African Botanical Diversity Network Report 9*. South African National Biodiversity Institute, Pretoria.

Hutchinson, J. 1946. *A botanist in South Africa*. Gawthorn, London.

Karsten, M.C. 1951. *The Old Company's Garden at the Cape and its Superintendents*. Maskew Miller, Cape Town.

Lighton, Conrad. 1960. *Cape Floral Kingdom*. Juta and Co., Cape Town.

McCracken, D.P. & McCracken, E.M. 1988. *The Way to Kirstenbosch*. National Botanic Gardens, Cape Town.

Mucina, L. & Rutherford, M.C. 2006. 'The Vegetation of South Africa, Lesotho and Swaziland'. *Strelitzia 19*. South African National Biodiversity Institute, Pretoria.

Mulder, C. 1988. 'Development Guide Plan: Kirstenbosch Botanic Garden'. Report to the Department of Public Works and Land Affairs, Pretoria.

Pauw, A. & Johnson, S. 1999. *Table Mountain – a natural history*. Fernwood Press, Cape Town.

Pearson, H.H.W. 'A National Botanic Garden'. *South African Journal of Science*, November 1910.

Pearson, H.H.W. 1911. *A State Botanic Garden*. The State of South Africa, Cape Town.

Pitt, B. & Boulle, T. 2010. *Growing Together: Thinking and Practice of Urban Nature Conservators*. SANBI Cape Flats Nature, Cape Town.

Raimondo, D., van Staden, L., Foden, W., Victor, J.E., Helme, N.A., Turner, R.C., Kamundi, D.A. & Manyama, P.A. (eds). 2009. 'Red List of South African plants 2009'. *Strelitzia 25*. South African National Biodiversity Institute, Pretoria.

Rycroft, H. B. 1980. *Kirstenbosch*. Howard Timmins Publishers, Cape Town.

Warner, B. & Rourke, J. 1998. *Flora Hercheliana: Sir John and Lady Herschel at the Cape 1834 to 1838*. The Brenthurst Press, Third Series, Johannesburg.

Index

Page numbers in italics indicate illustrations

A
Ableseius californicus 127
Abrahamse, Tanya 235
Acocks, John Philip Harison 60, *171*, 171
Adamson, R.S. 36
Adonis, Adonis 140, *141*
African Plants Initiative 36, 87, 95
Afrotemperate forest 17, 22, 48, 60, 114, 129, 218, *219*, 235
Agapanthus 106, 131, 148, 163
Agapanthus walshii 91
Agathosma ovata 121
Alexander, Henry 19, 20, 104
Allerton, Sam 192
Aloaceae 76, 100, 116, 125, 127, 217, 221, 229, 235
Aloe 76, 76, 116, 125, 127, 217, 221, 229, *229*, 235
Aloe arborescens 4, *22–23*, 39
Aloe broomii 76, 76
Aloe dichotoma *74–75*, 168, *168*, *206–207*, 213
Aloe ferox 76, 76, *100–101*
Aloe pearsonii 213
Aloe pluridens *34–35*
Aloe striata *100–101*
Amaryllidaceae 95, 230
Amaryllis 65
Ambrose, Barry 188, 189
Ammocharis 221
Anderson, Fay 89
Apocynaceae 125
Apodytes dimidiate 129
Araucaria 64
Archer, Joseph 212
Arctotheca 131
Arctotis 131, 148
Arctotis acaulis *142–143*
Arctotis hirsuta *142–143*
arum lily 148
Ashwell, Alice (Ally) 151, 154, 155, 163, 167
Attenborough, Sir David 231
Auge, Johannes Andreas 25

B
Babiana 231
Babiana framesii *231*
Baboon Spider *183*
Banks, Sir Joseph 37, 79
baobabs 125, 126, 127
Barker, Winsome 'Buddy' *43*, 94
Barkly, Sir Henry 26
Basson, Abraham 'Awie' *42*, 117
Basson, George *42*
Basson, William *42*
Batten, Auriol 89
Bauer, Ferdinand 82
Bauer, Frans 89, *89*
Baxter, W. Duncan 29, 33, *38*
Bayer, Bruce 212
Bean, Anne 153
Beattie, J.C. 28
berg winds 112, 114
Bergh, Kay 95, 102, 189, *189*
biodiversity 7, 12, 18, 58, 60, 62, 69, *70*, 71, 84, 85, 86, 95, 104, 105, 106, 107, 136, 138, 166, 167, 170–171, 172, 178, 180, 181, 182, 189, 217, 225, 230, 234, 235
biomes 26, 49, 62, *62*, 106, 180, 182, 213, 217, 221
Bird, Christopher 19, 20, 31
Bird's Bath 20, 30, *31*
Bolus, Harry 26, 27, 29, 43, 84, *84*, 93, 116, 152, 153, 186
Bolus, Louisa (née Kensit) 43, 152, *152*
Bolusanthus speciosus 217
Bond, Pauline 86
Bond, William 64, 168
Boomslang 182, *183*
Boophone 221
Boos, Frans 82, 140
botanical artists 82, 89–91
Botanical Research Institute *see* NBI
Botanical Society of South Africa 29, *38*, 45, 51, 102, 103, 104, 107, 116, 125, 138, 136, 153, 155, 159, 186, *187*, 189, 190, 194, 198, *203*, 208, 212, 222, 234, 236–237

bookshop 95, 199
 field guides 95, *95*, 163
Botanical Society Conservatory *105*, 125, 126–127, *126*, 130, *196–197*
Botha, Daan 102
Botha, Louis 28, 29, *29*, 189
Botha, Roelf 49
Boulle, Theresa 161
Bowie, James 82
Bowler, Dickie 120, 172
Brabejum stellatifolium 18, *19*
Bramwell, David 136
Brianhuntleya intrusa 212
Brown-veined White Butterfly 227
Brunia stokoei 69, 218
Bruniaceae 68, 69, 220
Brunsvigia 221
Brunsvigia bosmaniae *234*
Buddleja saligna 221
Bulbinella latifolia 114, *130–131*
Burchell, William J. 15, 16, *16*, 25, 76, 82, 109, 119, 167
Burchellia bubalia 86, *86*
Burkea africana 217

C
Caledon Wild Flower Show *144*, 145, 146
Camphor avenue *24*, 51, *236–237*
Cannomois grandis 65, *174–175*
Canthium inerme 129
Cape Batis 132, *132*
Cape Canary 132, *133*
Cape Dwarf Chameleon *179*
Cape Floral Kingdom 7, 12, 54, 57, 58, 59, 60, 64, 70, 71, 72, 73, 117, 119, 137, 120–121, 219
Cape Fox 182, *183*
Cape Grysbok 182, *183*
Cape Robin-Chat 132, *133*
Cape Spurfowl 132, *133*
Cape Sugarbird *70*, 132
Caracal 182
carbon dioxide 168, 169
Carols by Candlelight *184–185*
Carpobrotus edulis 149
Cartwright, Frank 33

Cassine peragua 129
Castle Rock *11*, 16, 20, *22–23*, 111
Casuarina 64
Cedrus atlantica 33
Celtis africana 229
Centre for Home Gardening 104
Chaplin, Peter 229
Chasmanthe 131
Chelsea Flower Show 49, 147, *147*
 gold medals 147
Chrysanthemoides monilifera 149
Church of the Good Shepherd 53
Cinnamomum camphora 225
Claudius, Heinrich 76
climate change 69, 107, 166, 167–168, 171, 178, 180
Clivia 121, 127, 131, 223
Clivia miniata 120
Clivia mirabilis 121
Cloete, Charles Duffy Henry 20
Codd, Leslie Edward Worstal 83
Company's Garden 25, *25*, 110
Compton, Robert Harold 12, 17, 28, 35, 36, *36*, 38, *38*, 43, 44, 45, *46*, 53, 93, 116, 118, 153, 160, 166, 186, 204, 208, 211, 236
Compton Herbarium 26, *43*, 47, 86, 88, 89, 93–95, *93*, 103, 104, *105*, 140, 145, 186
Condy, Gillian 89, 99, *99*
Cook, James 37, 78, 79
Cook, Mary Alexander 17
Cornelisson of Zevenhuysen, Leendert 17, 18, 22
Cotula 131
Cotyledon 221
Cotyledon orbiculata *100–101*
Cowling, Richard 64, 68
Crassula 213
Crassulaceae 100, 125
Crinum campanulatum 54
Crocodile River 222, *232*, 228, 229
Cunonia capensis 129, 220
Curtisia dentata 129
Cussonia paniculata 221
Cyanotis speciosa 226
Cycad Amphitheatre *32*, *174–175*

cycads 30, *31, 32, 49*, 128, 172–173, *172–173*, 223
   illegal trade 172, *173*

D

Dactyloctenium australe 223
Darting Cruiser Dragonfly *227*
Davidson, David 147
Davis, George 159
De Lange, Hannes 53, 176
De Villiers, Lord 28, 29, *29*, 31, 33
De Villiers Graaff, Sir David 29
De Winter, Bernard 44
Devils Peak 16, 112, *113*
Diaphanathe xanthopollinia 91
Diascia 231
Dicksonia antarctica 31
Dietes 131
Dimorphotheca 131
Dimorphotheca pluvialis 142–143, *214–215*
Diospyros whyteana 129
Disa uniflora 182, 219, 220
Donaldson, John 172, 173
Drege, Franz 82
Drosanthemum bicolor 11, 54, 125
Drosanthemum speciosum 11, 114, 125, 211
Drosera cistiflora 226
Duncan, Graham 131
Dyer, Robert Allan 44, *44*, 86, 216

E

Ecklon, Christian 82
Edith Stephens Wetland Park 159, *160*
Egyptian Goose 180
Eksteen, Dirk Gysbert 20, 48
Elandsberg Nature Reserve 145, *145*
Eloff, J.N. (Kobus) 50, *50*, 53, 54, 94, 138, 166, 176, 204
Encarsia formosa 127
Encephalartos 128
Encephalartos hirsutus 138
Encephalartos inopinus 173
Encephalartos woodii 172, *173*
endemism 28, 59, 71, 72
Englerophytum magalismontanum 217
Erica 42, 57, 65, 71, 79, 106, *119*, 120, 121, *121*, 124, 137, 138, 144, 220
Erica massonii 79, 89
Erica patersonii 82
Erica penduliflora 88

Erica regia 65
Erica turgida 120
Erica verticillata 120, *134–135*, 138, *139*, 140, *140, 141*
   re-discovery 140–141
   re-introduction 140–141
Ericaceae 42, 57, 60, 65, 68, 71, 79, 95, 106, 119, 120, 121, 124, 137, 144, 220
Esterhuysen, Elsie 84, *84*
Euclea crispa 221
Eucomis 131
Eucomis autumnalis 163
Euphorbia cooperi 34–35
Euphorbia dregeana 210
Euphorbia ingens 39
Euphorbiaceae 125
Euryops 131
Euryops pectinatus 22–23
extinction 59, 70, 71, 72, 138, 140–141, 172, 173

F

Fabaceae 68
Fata, Arthur, 'Wild Seedpod' 192
Felicia 131
Felicia australis 232–233
ferns 30, 59, 71, 72, 127, 160, 188, 213
Fernwood Buttress *13*, 111, 112, *113, 118*
Ficus macrocarpa 225
fire 48, *66–67*, 68, 69, 114, 118, 124, 141, 145, 161, 166, 169, 176, 219
floral kingdoms 58, *58*
Foden, Wendy 168
Fredericks, John *42*
Free State NBG 208, 210, 221–222, *221*
Freesia 65, 144, 148
Fynbos Biome *22–23*, 26, 60, *63*, 64, 66, 68, 71, 120
Fynbos Biome Project 60, 64

G

Gardenia globosa 91
Gazania 131
Gazania rigida 214–215
Geissolomaceae 68
Geissorhiza 144
Geissorhiza radians 130
geophytes 65, 68, 131, 213, 221, 230
George VI, King *46*
Gladiolus 65, 89, 144, 148, 163
Gladiolus caryophyllaceus 148

Gladiolus virgatus 64
Gloriosa 131
Goldblatt, Peter 86, 87, 231
Gold Fields Environmental Education Centre 103, 153
   education officers 154–155
Gondwana 69, 111, *111*, 182
Green-veined Charaxes Butterfly *227*
Grewia occidentalis 221
Grubbiaceae 68
Gunn, Mary 83

H

Hadeda Ibis 178
Haemanthus 221
Haemanthus coccineus 77
Hall, H.L. 222
Halleria lucida 129
Hantam NBG 74, 131, 210, 230–234, *230, 232–233*
Harold Porter NBG 208, 210, 218–220
Harvey, William Henry 37, 86, 87, *87*
Heatlie, Charles 212
Helmeted Guineafowl 132, *132*, 180
Hemimerris 231
Hermann, Paul 57, 76
Herschel, John William Frederick 89
Herschel, Margaret Brodie 89
Hesperantha pauciflora 232–233
Hesperantha viginata 130
Heurnius, Justus 76
Heywood, Vernon 136
Hill, Leslie 178, *189*
Hilton-Taylor, Craig 72
Hirpicium integrifolium 212
Hitchcock, Anthony *134–135*, 137, 140
Hoffman, Timm 167
Honig, Marÿke 162, 163
horticulture 37, 50, 53, 79, 106, 124, 137, 139, 141, 158, 163, 166, 176, 180, 199, 204, 209, 210, 211, 212, 213
hot spots 57, 69, 70, 71, 72, 172, 233
Hoy, Sir William 211
Hudson, Raymond 147
Huntley, Brian John 98, *98*, 106, 188, 204, 210
Huntley, Merle 90
Hyacinthaceae 230
Hypoxis 221

I

Ilex mitis 129, *174–175*, 219
indigenous flora
   as invasive aliens 148
   conservation 136
   exports 148–149
   horticulture 27, 118
   preservation 28, 136
invasive aliens 22, 26, 48, 98, 118, 129, 148, 149, 166, 180, 225
Iridaceae 65, 68, 95, 230, 231
Ischyrolepis subverticillatus 113
Isoetes capensis 160
Ixia 65, 79, 144, 148
Ixia viridiflora 79

J

Jacobs, Andrew 117, *117*
Jacobs, Clive 117, *117*
Jacobs, Dennis 117, *117*
Jacobs, Freddie 117, *117*
Jamesbrittenia bergae 94
Johns, Muriel E 152, *153*
Josephus, Nicholas *42*

K

Karoo Desert NBG 49, 207, 208, 210, 211–215
Keeble, Sir Frederick 225
Kew Royal Botanic Garden 26, 27, 30, 33, 36, 37, *37*, 79, 82, 87, 89, 99, 118, 119, 138, 140, 144, 148, 168, 191, 208, 219
Kgope, Barney 168, *169*
Kidd, Mary Maytham 89
Kigelia africana 91
Kiggelaria africana 129
Kirsten, Willem Hendrik 19
Kirstenbosch Botanical Garden 80–81
   aerial photograph *204–205*
   and the community 158–159
   annual reports 186
   bell tower *46*, 47, 186
   Biennales 76, 90–91
   biological controls 127
   birds *132–133*
   builders *42*
   buildings 186, *187*
   climate 112
   'compost factory' 116, *124*
   concert stage 190
   Dell 20, 30, 31, *31*, 106, 116
   design 116, 118
   'Enchanted forest' *120*, 121
   entrance fees 194

environmental education 152, 153, 154–155
Erica Garden *119*, 120
fauna 182–183, *182–183*
financial independence 204
Fragrance Garden 163
freeway plan 47
funding 29, 104, 158, 185–205, 210
Garden Centre *199*, 202
Garden of Extinction *137, 155*, 163
Garden of Names *162*, 163
Garden of Useful Plants 163
geology 111
glasshouses 104, 136
gold medals 147
gorilla statues 192
Hampton Court Palace exhibit *144*
image 204
infrastructure 102, *104*, 198
invasive aliens removal 48
irrigation system 116
logo *107*
maps *38, 116, 187, 236–237*
Marquee Lawn 20, *194*
Moyo Kirstenbosch Restaurant *194*
name 19
Old Mutual Conference Centre *193*
Otter pond *49*
outreach greening programme 158, 159, *159*
parking areas 188, *188*
plans *20*, 51, 53
plant labels *138*
profit 185, 194, 198, 199
Protea Garden 42, 102, 103, *119*
rainfall 112
ralli-cart 43, *43*
research 50, 166
reservoir *115*, 116, *117*
restaurant 104
Restio Garden 121, *176–177*
Sanlam Hall 153, *155*
science programme 166, 167, 181
Sculpture Garden *190*, 191
seed banking 138
Silver Jubilee 1938 45, *45*
site 12, 110
soil 124
 mulch 124
staff 36, 43, 50, 52, 53, 106, 117, *117*, 203, 204

Summer Sunset Concerts 190, *200–201*, 202
Tea House 104, 186, *202*
theme gardens 163, 191, 192
transport 43
visitors 193, 194, 195
 expenditure 198
Visitors' Centre *105, 198*
volunteers 154, 155
Water-wise Garden 163
wedding venue 192
Kirstenbosch bus *150–151*, 158, 159
Kirstenbosch Development Campaign 102, 103, 106, 189, 203, 204
Kluge, Johan 223
Knysna Warbler 132, *132*
Kogelberg *57, 59, 60–61*, 69, 77, 84, 218, 219, 220
Kotze, Deon 140
KRC (Kirstenbosch Research Centre) 94, 103, 104–105, *164–165*, 168, 172, 179, 180
Leslie Hill Molecular Laboratory 178
Molteno Library 104
Krieger, Frank *42*
Kruger, Fred 60
KwaZulu-Natal NBG 224–225
Kwelera NBG 235, *235*

L
L'Écluse, Charles de 77, *77*
L'Obel, Mathias de 77
Lachenalia 65
Lady Anne Barnard's Bath 20
Lampranthus amoenus 11, *114, 125*
Lampranthus hoerlenianus 211
Lampranthus spectabilis 114
Lanariaceae 68
Lapeirousia 231
Le Roux, Philip 203, 204
Leendertsbos 17, *17*, 19, 116
Leucadendron 64
Leucadendron argenteum 65, *68*, 119
Leucadendron tinctum 170
Leucospermum cordifolium 55, *109, 118, 226*
Leucospermum erubescens 65, *171*
Leucospermum oleifolium 110
Leucospermum reflexum var. luteum 118
Levyns, Margaret (née Michell) 36, *36*

Lewis, Dylan 192, *193*
Lichtenstein, Martin 82
Lincoln, Thalia 89
Linnaeus, Carolus 57, 76, 78, 85, *85*, 86
Lithops 127
Logan, James D. 211
Lower Contour Path 17
Lowveld NBG 208, 210, 222–223

M
MacGregor, Neil 230
MacKinnon, Kathy 149
Maclear's Beacon *13, 46*
MacOwan, Peter 25, 26, 78, 185
Mandela, Nelson 97, 98, *98*, 99, *99*, 154, *155*, 198, 229
Manning, John 86, 87, 89, 145, 231
Marais, Jack 46, 48, 129
Marloth, Rudolf 28, 58, *58*, 59, 84, 114, 182
Marriott, W.E. 224
Marsh Terrapin 182, *182*
Masson, Francis 37, 75, 78, 79, *79*, 89, 99, 148, 168, 219, 220
Mathews, Joseph William 12, 16, 30, *30*, 37, 38, *38*, 39, 46, 98, 116, 125, 131, 149, 166
Mathews Rockery *34–35, 38, 39*, 118, 125
McGregor, Alec 190
Mclean, 'Driver' David *42*, 117, *117*
Merriman, John X. 28
Mesembryanthemaceae 68, 125, 152, 213
Midgley, Guy 167, 168
Millennium Seed Bank 36, 138, 140
Mimetes argenteus 170
Mimetes hottentoticus 220
Mkefe, Xola *154*, 158
molecular biology 178, 179, 181
Monkey Beetle 231
Moraea aristata 72
Moraea calcicola 73
Moraea gigandra 73
Moraea loubseri 73
Moraea tulbaghensis 73
Mother-of-Pearl Butterfly *227*
Mountain Ghost Frog 182, *183*
Mountain Pride Butterfly 182, *183*
Mountain Reedbuck 229
Mullins, Mary 102, *189*, 190
Myrsine Africana 129

N
Natal Botanic Garden 208, 224, 225, 228
Natal Herbarium 44, 98, 216
National Herbarium 28, 44, 45, 50, 84, 86, 93, 95, 98, 102, 208, 216, 217
Naude, Meiring 50
NBG (National Botanical Gardens) 49, 50, 53, 54, 94
 leadership teams 209, *209*
 logo *107*
 South Africa 208
  map *208*
NBI (National Botanical Institute) 46, 50, 54, 98, 99, 102, 104, 105, 106, 136, 138, 153, 154, 158, 159, 166, 167, 171, 188, 210, 230
 logo *107*
Nerine 221
Nicholas, James *42, 43*
Nieuwoudtville 121, 131, 168, 231, 234
Niven, James 82, 83
Nivenia stokoei 83
Nursery Ravine 48, 114, 129

O
Ochna pulchra 217
Ocotea bullata 129
Olea capensis 129
Olea europaea 221
Olinia ventosa 129, 220
Olive Thrush 132
Oliver, Ted 88, 140
Orange-breasted Sunbird 132, *139*, 141
Orothamnus zeyheri 88, *226*
Ornithogalum 65, 144
Ornithogalum dubium 131
Osteospermum 131
Oxalis obtusa 232–233

P
Pachypodium namaquanum 210
Painted Reed Frog 178, *178*, 180
Pappe, Ludwig 25, *25*, 26, 86, 88
Parfitt, Yvonne Eustasie 209
Paterson, William 82, *82*, 83
Pearson, Henry Harold Welch 11, 12, 15, 16, 21, 25, 27, *27*, 28, 29, 30, 31, 32, 33, *33*, 36, 37, 44, 48, 49, 54, 87, 104, 110, 127, 128, 131, 137, 144, 149, 152, 163, 165, 166, 172, 178, 180, 203, 204, 207, 208, 211, 213, 235

Index 239

Pearson House 50, 102, *103*
Pearson Memorial Hostel 186, *188*, 209
Pelargonium 37, 79, 137, *148*, 148, 213
Pelargonium peltatum 94
Peter, Zwai 159
Petersen, Stella 154, *154*
Phillips, Sir Lionel 28, 29, *29*, 33, 46, 47
Phytophthora cinnamomi 119, 120, 124
Pillans, Neville 16, 211
Pitt, Bridget 161
Platanus x acerifolia 225
Plectranthus 30, 89, 106, 121, *122–123*
Plectranthus ecklonii 122–123
Pleistocene 68, 111
Podocarpus 64, 69
Podocarpus falcatus 174–175
Podocarpus latifolius 18, 129, *174–175*
Pole Evans, Illtyd Buller 28, *44*, 44, 83, 208, 216
Polygalaceae 68
Porter, Harold 218
Portulacaceae 125
Prance, Sir Ghillean 138
Pretoria NBG 216–217
Prionium serratum 219
Protea 42, 49, 55, 57, 64, 71, 86, 102, *103*, 106, 116, *118*, *119*, 119, 120, 121, 124, *138*, 144, 163, 178, 220
Protea cynaroides 86, 119, *128*, *171*, 220, *220*
Protea neriifolia 77, *77*, 91
Protea Atlas Project 170–171, *170–171*
Protea Village 52, *52*, 53
 forced removals 52
Proteaceae 42, 49, 57, 60, 64, 65, *65*, 68, 69, 71, 86, 94, 95, 106, 116, 118, 119, 120, 121, 124, 138, 144, 163, 170, 178, 220
Pterocelastrus tricuspidatus 19

R

Raimondo, Domitilla 71
Rapanea melanophloeus 129, 220
Raven, Peter 136
Rebelo, Tony 170
Red List of South African Plants 71, *71*, 137, 141
Restionaceae 53, 57, 60, 65, *65*, 69, 71, 119, 121, 176, *176–177*, 220

Retzia capensis 69
Rhamnaceae 68
Rhodes, Cecil John 16, 20, *21*, 24, 47, 154, 158
 properties *21*
Rhodes Drive 46, 48, 51, 102, *105*, 186, *204*
Richardson, David 64
Ridley, George Herbert 16
Romulea 231, *231*
Romulea monadelpha 130, 231, *231*
Romulea sabulosa 231, *231*
Rondevlei Nature Reserve 135, 141
Roridulaceae 68
Rosaceae 68
Ross, Sue 190, 201
Rouget, Mathew 140
Rourke, John 47, 83, 89, 94
Roux, Koos 95
Ruschia caroli 212
Rutaceae 68, 121, *121*
Rutherford, Michael 62
Rycroft, Hedley Brian *42*, 46, *46*, 47, 48, 49, 50, 51, 53, 84, 94, *117*, 147, 166, 208, 210, 221, 222, 228

S

SABONET (South African Botanical Diversity Network) 87, 88, 95, 162, 210
SANBI (South African National Biodiversity Institute) 12, 71, 72, 83, 87, 99, 106, 107, 136, 137, 138, 159, 161, 170, 171, 172, 173, 181, 203, 209, 225, 235
 logo *107*
 partner organisations 106, 180
Schimper, Andreas Franz 59
Scholl, George 82, 140
Schotia brachypetale 122–123
Schreiner, W.P. 28
Searsia lancea 221
Senecio 131
Senecio crassulifolius 100–101
Senecio elegans 142–143
Sherwood, Shirley 89, 90
Silver Tree Trail 17
Sisulu, Walter 229
Skeleton Gorge 48, 84, 129
Slug-eating Snake 127
Small Grey Mongoose 182, *182*
Smith, Sir Frederick 28
Smith, Gideon 87
Smith, Robbie 38
smoke-based germination 53, 176
smoke primer 176

Smuts, Jan Christian 44, 45, *45*, *46*, 145, 212, 216
Sombre Greenbul 132, *133*
Sonder, Otto Wilhelm 87
South African Association for the Advancement of Science 12, 28
South African College 16, 25, 27, 28, 29, 37, 84, 93
Sparaxis 144
Sparaxis elegans 65
Sparrman, Anders 75, 78, *78*, 79, 167
Sparrmannia africana 86, *86*
Spotted Eagle-Owl 132, *132*
Starke, Laetitia 153
Stearn, William 89
Steiner, Kim 231
Stephens, Edith 36, *36*, 154, 159, 160, *160*
Stokoe, Thomas Pearson 84, *84*
Stone Cottages 51, 186, *187*
Strelitzia juncea 34–35, *91*
Strelitzia nicolai 90
Strelitzia reginae 79, 148
Strelitzia reginae 'Mandela's Gold' 97, 99, *99*
Streptocarpus 30, 121, 127, 137, 148
Streptocarpus kentaniensis 82
succulents 35, 39, 100, 104, 125, 126, 127, 152, *207*, 212, 213, 217, 229
Swift, John 234
Swift, Kirsten 234

T

'table cloth' 114
Table Mountain Ghost Frog 182, *183*
Table Mountain National Park 13, 72, 117
Tarr, Brian 225
Taxodium distichum 225
taxonomy 85–86, 87, 93, 94, 95, 166, 178, 181
Thick-toed Gecko 127
Thiselton-Dyer, Sir William Turner 26, 27, *27*, 37, 87, 208
Thoday, David 36
Thomas, Vicki 89
Thompson, Jorgen 234
Thompson, Julian Ogilvie 189
Thorns, Frank 37, 46, 118
Threatened Plants Research Laboratory 50, 53, 176
threatened species 72, 120, 131, 136, 137, 140, *141*, 148, *162*, 172, 173, 192, 235

Threatened Species Programme 72, 137–138, 163
Thudichum, Jacques 212
Thunberg, Carl Peter 75, 78, *78*, 79, 86, 167, 168, 230
Thunbergia natalensis 78, 86, *86*
Tolley, Krystal 178, 179
traditional medicine 121, *121*, 163, 172
Tripteris oppositifolia 210
Tulbagh, Rijk 76, 85

U

Ursinia 131
Ursinia anthemoides 142–143

V

Van der Stel, Simon 76
Van Jaarsveld, Ernst 89, 104, 121, 126
Van Riebeeck, Jan 17, 18, 25
Van Riebeeck's Hedge 18, *18*
Van Zyl, Gideon Brand 212
vegetation map *62*
Velvet Worm 182, *182*
Verreaux's Eagle *228*, 229
Virgilia capensis 220
Virgilia oroboides 129
Vitaceae 125
Von Mueller, Baron Ferdinand 135

W

Walter Sisulu NBG 208, 210, 228–229
 Geological Garden 229
Ward-Hilhorst, Ellaphie 89
Watsonia 65, 131, 144, 148
Welwitschia mirabilis 7, 27, *27*, 127
Werner, H.F. 37, 46, 118
Widdringtonia 69
Widdringtonia cedarbergensis 174–175
wild flower shows 144, 145
Willis, Christopher 210
Winter, John 37, 46, 99, 106, 118, 120, 121, 172
Witpoortjie waterfall *228*, 229
Wood, John Medley 44
Wyse-Jackson, Peter 136

Z

Zeyher, Carl 25, 26, 82, 88
Zimbabwean stone sculptures 191, *191*